Humanism and its Discontents

Paul Jorion
Editor

Humanism and its Discontents

The Rise of Transhumanism and Posthumanism

Université Catholique de Lille 1875

Région Hauts-de-France

palgrave macmillan

Editor
Paul Jorion
Department of Ethics
Université Catholique de Lille
Lille, France

ISBN 978-3-030-67003-0 ISBN 978-3-030-67004-7 (eBook)
https://doi.org/10.1007/978-3-030-67004-7

This Palgrave Macmillan imprint is published by the registered company Springer Nature Switzerland AG.
The registered company address is: Gewerbestrasse 11, 6330 Cham, Switzerland

Contents

Notes on Contributors

Salomé Bour holds a PhD in Philosophy. She is an associate researcher at the Centre for Contemporary Ethics (Lab. Epsylon, Université Paul-Valéry, Montpellier). She is also a secondary education philosophy teacher. Her works focus on extropianism, the philosophy of transhumanism and ethical issues raised by the transhumanist movement.

Marius Dorobantu is a postdoctoral research associate in the Theology and Science Department at the Vrije Universiteit Amsterdam (Netherlands). He is working on a project within the "Diverse Intelligences" initiative funded by the Templeton World Charity Foundation. He holds a PhD in Ethics from the University of Strasbourg (France), with a thesis titled "Theological Anthropology and the Possibility of Human-Level Artificial Intelligence: Rethinking Human Distinctiveness and the Imago Dei".

Francis Heylighen is a research professor at the Vrije Universiteit Brussel, where he directs the Leo Apostel Centre for Interdisciplinary Studies and the Global Brain Institute. He investigates the self-organization and evolution of complex systems from a cybernetic perspective.

Hélène Jeannin holds a PhD in Sociology and an MD in Communication and French Literature. She works as a sociologist on a variety of forward-looking topics in the Social and Human Sciences Research Department (SENSE) of Orange Labs in Châtillon.

Paul Jorion is an associate professor in the Department of Ethics, Technology and Transhumanism at the Université Catholique de Lille. He holds a PhD in Social Anthropology and an MA in Sociology. He is a practising psychoanalyst. He was a contributor to the Connex artificial intelligence project of British Telecom. For 18 years, he was also a pioneer in financial algorithms in the USA, the UK, France and the Netherlands.

Susanna Lindberg is Professor of Continental Philosophy at the University of Leiden, the Netherlands. She holds a PhD and a Habilitation à diriger les recherches in Philosophy. She is specialized in German idealism, Heidegger and contemporary French philosophy. She has studied and worked in Finland and France.

Stefan Lorenz Sorgner is Professor of Philosophy at John Cabot University in Rome, is director and co-founder of the Beyond Humanism Network, a fellow at the Institute for Ethics and Emerging Technologies (IEET), a research fellow at the Ewha Institute for the Humanities at Ewha Womans University in Seoul and a visiting fellow at the Ethics Centre of the Friedrich-Schiller-University in Jena.

Alexander Thomas is a part-time PhD student at the University of East London. His research focuses on the ethics of transhumanism in the era of advanced capitalism. He also leads the Media Production degree course at the University of East London and is a multi-award winning film director and screenwriter.

Clément Vidal holds a PhD in Philosophy with a background in logic and the cognitive sciences. In 2014, he authored *The Beginning and the End: The Meaning of Life in a Cosmological Perspective*. He is eager to tackle big questions, bringing together areas of knowledge such as cosmology, physics, astrobiology, complexity science or evolutionary theory.

Introduction: Humanism and Its Discontents—The Rise of Transhumanism and Posthumanism

Paul Jorion

Humanism was defined by two pronouncements in the *Book of Genesis*: Man was made to God's image and God had entrusted Man with the dominion of all creatures.

The *Homo Imago Dei* part of the Christian definition went into a crisis in the eighteenth century with the Enlightenment and was henceforth dropped for all practical purposes.

The rise of different types of anti-humanism or, one should rather say, varieties of neo-humanism, such as Nietzsche's superhumanism, posthumanism, transhumanism and del Val and Sorgner's metahumanism (2010), finds its origin in the crumbling of humanism over the centuries due to doubts accumulating around the Western view of humankind rooted in the *Homo Imago Dei*, and of its follower: the *perfectible* Man of an Enlightenment enamoured with never-ending progress.

P. Jorion (✉)
Department of Ethics, Université Catholique de Lille, Lille, France
e-mail: paul.jorion@univ-catholille.fr

P. Jorion (ed.), *Humanism and its Discontents*,
https://doi.org/10.1007/978-3-030-67004-7_1

Part I. Humanism on the Wane

When in the mid-1960s Michel Foucault evoked our representation of Man as possibly on the wane—"It is comforting, however, and a source of profound relief to think that man is only a recent invention, a figure not yet two centuries old, a new wrinkle in our knowledge, and that he will disappear again as soon as that knowledge has discovered a new form"—it was not entirely clear from the contents of *The Order of Things* (1966) why this remark would come as a conclusion of sorts at the end of the book.

There were two clues however. The first was that the French philosopher's book was aiming at being an archaeology of the concepts of the contemporary social sciences ("sciences humaines") and was examining how different *épistémès* succeed each other over the ages, *épistémès* being worldviews but restricted in their scope as highbrow ways of depicting the world, and therefore as tools for knowing and thinking. An *épistémè*, Foucault wrote, "defines the conditions of possibility of all knowledge". The second clue was in the "not yet two centuries old", signalling that the "man" in question was undoubtedly that of the Enlightenment.

The concept of Man which is currently ours, hints Foucault, is likely to evolve in the same way as all depictions have over the ages, such a process being illustrated in his earlier and later works, when he scrutinised the shifting assessment within our culture of madness, and what we have regarded as *normalcy* in our sexual behaviour, where he follows in the footsteps of his master Georges Canguilhem who reflected in his time on *The Normal and the Pathological* (1943). One grasps under that light where his doubts on the perennial character of a particular representation of humankind arose from. What remains more mysterious is why "it is comforting […] and a source of profound relief to think that" the Enlightenment's Man would soon be gone.

One thing is sure then from the scope of *The Order of Things* with its focus on the Enlightenment, complemented by the confirmation provided by the unambiguous "not yet two centuries old" specification: Foucault's starting point, the view of Man possibly on the wane, was not that of the Christian *Homo Imago Dei*, "Let us make Man in our image, after our likeness" (Genesis: 1.26), but that of what was by then its successor, that of a human race whose nature is dynamic and mouldable, as its horizon is that of *perfectibility*, indefinitely unfolding, as Jean-Jacques Rousseau was first to underline. The "Solitary Walker » wrote indeed" But when the difficulties surrounding all these questions would leave some room for argument about this difference between man and animal, there is another very specific quality that distinguishes them, and on which there can be no dispute, it is the ability to perfect oneself" (Rousseau III: 142), a notion that Nicolas de Condorcet would further develop, while strategically assigning its conception to his mentor Turgot instead of Rousseau, the inspirer of his political adversaries Robespierre and Saint-Just.

From inception, the concept of Man in the image of God had shown frailties. What is indeed precisely the extent of the likeness: does God display any of our shortcomings? Does he need to eat and drink? To pee and poop? Most unlikely features for a perfect entity. But if not, what are the exact constraints on us rooted in the supposed God's likeness?

Two periods of disarray in the history of Western humanism can indeed be traced, the first being the demise of *Homo Imago Dei* and the second being the crisis of the Enlightenment's version of humanism, that of a humankind left to its own devices, having discarded the first prong of the biblical message, "And God said, Let us make Man in our image, after our likeness", while still remaining attached to the second, this time proactively: "… and let them have dominion over the fish of the sea, and over the fowl of the air, and over the cattle, and over all the earth, and over every creeping thing that creepeth upon the earth" (Genesis: 1.26).

Christian theology is still tightly linked to *Homo Imago Dei* and as Marius Dorobantu underlines in his contribution to this volume: "Strong Artificial Intelligence and theological anthropology: one problem, two solutions", it has the greatest difficulty cutting the cord in a context where more and more of the defining traits of humankind are now shared by machines. Dorobantu reviews in his chapter a number of options open to Christian theology, all equally unsatisfactory. One is denying in retrospect the importance for the Christian faith of the *Homo Imago Dei*. Another resides in displacing the stress from the human nature to the relationship between Man and his Creator. As he emphasises, however, the very essence of the Turing test for Artificial Intelligence lies not so much for the machine in emulating likeness in appearance with Man but in manners, that is, in the quality of rapport established between human beings.

The final option, in the Hegelian tradition, is setting *Homo Imago Dei* not as a design having presided to Man's creation but as an aim for human beings over the ages. That would mean that even future super-intelligent machines, should such a goal be indifferent to them, might still lack what it takes to truly be in the image of God.

The sudden realisation that "God is dead" or the slow realisation that "God has never been around" led to two types of attitudes, both having been reported: first, that of Dostoyevsky's Ivan Karamazov, according to whom "If there is no immortality, there is no virtue", and Nietzsche's, amounting to "Then all is allowed, let us be merry and revel in a permanent Dionysian bacchanal!"; and, second, that of the Enlightenment—in the absence of an all-pervasive God, the responsibility towards the universe is thence entirely ours. With this latter attitude, with the constraints attached to Man being in the image of God having been entirely relieved, and the option of the infantile response "Then let's be wild!" having been discarded, Man is free to improve himself, in what he determines—through introspection—as being his deficiencies or other failings, in his body and soul.

But in the same way as the Enlightenment's thinkers had become suspicious of the *Homo Imago Dei*, intellectuals of Foucault's generation had developed doubts about a limitlessly perfectible Man. The reason was—instead of Heideggerian qualms about technology's "enframing"—a realisation in the style of Gunther Anders' that the twentieth century was the time when the devastating possibility that technology would wipe out humankind had become inescapable. Whatever Foucault may have had in mind, let's remember that *The Order of Things* was published in 1966, only four years after the Cuban Missile crisis of 1962, a time when the likelihood of ultimate doom was, for good reasons, impossible to dispel, even if Foucault mentions a more benevolent source for Man's "disappearance": "as soon as […] knowledge has discovered a new form", that is, a conceptual vanishing solely instead of proper annihilation.

If technology in the shape of applied science—to be distinguished from the empirical technology based on trial and error of the pre-Renaissance period—was capable of being the tool for perfectibility, it had become clear that it could as well wipe us out. There lay undoubtedly the germ of the major backlash and self-questioning against technology that can now be observed. Of much less importance in the drop in prestige of science is the possible reproach, and to technology in its capacity of being science's operating arm in everyday life, that it has failed at disproving once and for all the existence of God, a task once regarded by its proponents as within its reach, and transpiring in Victor Hugo's anecdote of when Napoleon complained about the absence of God in Pierre-Simon Laplace's *Mécanique céleste*—"How, you make the whole system of the world, you give the laws of all creation, and in your whole book you do not speak once about the existence of God!"—the latter had allegedly retorted, "Sire, I had no need for that hypothesis" (Hugo [1847–1848] 1972: 217).

Let us remember though what precisely Rousseau understood under that word of "perfectibility" lest we turn to it naively as a simple placeholder for whatever happened to us humans since we, as *Silvatici* at the time, left the woods, and forget about the ambivalences of such a disposition for the human race, as the illustrious Citizen of the Geneva Republic had himself been quick to emphasise. So much so that we can read in the reflection he devoted to *perfectibility* a warning against the harms of today's world and the perils that we are currently facing.

Let us therefore examine carefully Rousseau's pondering on *perfectibility*. Unless stated otherwise, all quotations are borrowed from his *Discourse on the Origin and Basis of Inequality Among Men* of 1755.

First of all, what is an animal?

> I see in every animal nothing but an ingenious machine, to which nature has given senses in order to wind itself up, and to guarantee itself, to a certain extent, from everything that tends to destroy or disturb it. (Rousseau III: 141)

And what is a human?

> Considering him (…) as he must have come out of the hands of Nature, I see an animal less strong than some, less agile than others, but on the whole, the most advantageously organised of all. (Rousseau III: 134–135)

"Most advantageously organised of all", as on the one hand unspecialised: accumulating the instincts other animals share between them, and on the other hand omnivorous: finding his sustenance in a wide range of foods where other species pick according to their more restricted likings.

> (Humans) rise to the level of the instinct of animals, with the advantage that each species has only its own instinct, and that man, perhaps having none of his own, appropriates them all, also feeds on most of the various foods that the other animals share, and consequently finds his subsistence more easily than any of them can do. (Rousseau III: 135)

The ease with which humans find their food must make them the most indolent of all animals, and this is what Rousseau, the "Solitary Walker", believes indeed that he's observed:

> I see him satiating under an oak tree, quenching his thirst at the first stream, finding his bed at the foot of the same tree that provided him with his meal, and with that, all his wants supplied. (ibid.)

It should be noted that this notion of humans as unspecialised creatures is reminiscent of the fable that Epimetheus tells in Plato's *Protagoras*. Due to inadvertence, Man has been left without qualities, which forces Prometheus to make up for such a lack by drawing on a fund of qualities which had not been retained for distribution among animals: as a gift to Man, he steals fire from Hephaestus and Athena (Plato, *Protagoras* 320c–322a). Rousseau sees here something entirely different: Man fills that lack through addition: adding up all various instincts to make up his own, adding up all varieties of food to ensure his own sustenance.

And this is precisely where perfectibility comes into play:

> But, when the difficulties surrounding all these questions would leave some room for dispute about this difference between man and animal, there is another very specific quality that distinguishes them, and on which there

> can be no dispute, it is the ability to perfect oneself, a faculty which with the help of circumstances, successively develops all the others, and resides among us both in the species, and in the individual, instead of an animal being, after a few months, what it will be all its life, and its species, after a thousand years, what it was the first year of those thousand years. Why is Man alone liable to grow into a dotard? Is it not that he thus returns to his primitive state, and that, while the Beast, who has acquired nothing and has nothing to lose either, always remains with his instinct, man losing through old age or other accidents, all that his perfectibility had made him acquire, thus falls back lower than the Beast itself? (Rousseau III: 142)

There is, however, with perfectibility a downside to what might seem at first sight an undeniable advantage:

> It would be sad for us to be forced to agree that this distinctive and almost unlimited faculty is the source of all man's misfortunes; that it is that faculty that draws him, over time, from his original condition, in which he would spend peaceful and innocent days; that it is that faculty that, over the centuries, has given birth to his enlightenments and his errors, his vices and his virtues, making him in the long run the tyrant of himself and of Nature. (Rousseau III: 142)

As can be seen here, *perfectibility* is the worm in the fruit, the faculty which, once in motion will, through its own momentum, draw Man out of the State of Nature, and will then preside over the succeeding ages of the human race.

Perfectibility constitutes the delicate joint where History connects with Nature. History is from early on inscribed in Nature because a dormant History constitutes one of the elements of "Natural Man's" essence. Development written in potentiality, prepared and ready in the seed, is capable of building up the whole plant to come, but the command for the process to begin needs to come from above, from another facet of the very same Nature, in motion this time rather than still.

On such dormancy, Claude Lévi-Strauss wrote the following in *From Honey to Ashes*:

> But the 'dormancy' of the seed, that is, the unpredictable time that will pass before the mechanism is triggered, is not due to its structure, but to an infinitely complex set of conditions that summon the individual history of each seed and a variety of external influences. The same is true for civilisations. Those we call primitive do not differ from others in their mental equipment, but only in that nothing in any mental equipment whatever, ever thought that it should deploy its resources at a certain moment and exploit them in a certain direction. (Lévi-Strauss 1966: 408)

> Let us conclude, writes Rousseau in the *Discourse*, that wandering in the forests without industry, without speech, without home, without war, and without connections, without any need of his fellow men, as well as without any desire to harm them, perhaps even without ever making any difference between them individually, the Savage man subject to few passions, and self-sufficient, had only the feelings and lights proper to this state, that he felt only his true needs, looked only at what he thought it was in his interest to see, and that his intelligence made no more progress than his vanity. If by chance he made some discovery, he could not communicate it, all the more so because he did not even recognise his own children. Art perished with its inventor; there was neither education nor progress, generations multiplied uselessly; and each starting always from the same point, centuries passed in all the crudeness of the first ages, the species was already old, and man always remained a child. (Rousseau III: 159–160)

Similarly, in his *Letter to Philopolis* (the naturalist Charles Bonnet [1720–1793]), Rousseau writes:

> Since you claim to criticise me through my own system, please remember that in my opinion society is as natural to the human species as decrepitude is to the individual and that there must be arts, laws, governments to peoples as there are crutches for the elderly. The whole difference is that the state of old age derives from the nature of man alone and that that of society derives from the nature of the human race, not immediately as you say, but only as I have proved, with the help of certain external circumstances which may or may not be, or at least come sooner or come later and therefore accelerate or slow down progress. (Rousseau III: 232)

What most obviously results from this is that Man could only remain in the State of Nature for a spell of time that would come to an end sooner or later. Then he would become sociable, whether sociability was inscribed in him or in the Nature around him: "Men's associations are to a large extent the work of accidents of nature: the particular deluges, the extravasated seas, the eruptions of volcanoes, the great earthquakes, the fires started by lightning which destroyed the forests…" (Rousseau n.d.).

We fully understand then Rousseau's dismay when he believes he has recognised the Natural Man in the orangutan that distant travellers describe: on the one hand, he asserted that Man (out of his own nature) was to remain in the Natural State, but on the other hand, the way that nature is currently to our eyes offers little to suggest (due to climate change, decline of soil fertility etc.) that the Natural Man could somehow have survived in his pristine state without his virtual faculties having been awakened by the stimuli that the turmoil and upheavals of external circumstances bring with them.

The loss marking the passage of *Man from the nascent society* to the *policed Man* ("The nascent society gave way to the most horrible state of war" [Rousseau III: 176]) may look like a benefit, as it displays Man developing his potential faculties through *perfectibility* and moving towards his perfection but the truth is the decrepitude of the species. The benefits of budding Reason and self-esteem were accompanied by the emergence of a depraved narcissistic self-love. No doubt, those contribute to ensuring the conservation of the individual, but natural pity, the species' benevolent warden, has faded away; the individual prospers, yet his prosperity is but a sham as at the same time the species as such is ensnared in decay.

Therefore, the more productive way of looking at the emergence of our four recent brands of anti- or neo-humanism (superhumanism, posthumanism, transhumanism and metahumanism) is likely to be to first notice the disarray arising in the Renaissance about the early humanism of the *Homo Imago Dei*, then observing humanism in its Enlightenment's *perfectibility* form sharing a similar fate "not yet two centuries" later—truly in the exact shape that Rousseau had anticipated, and see then in them a variegated set of four attempts at reconstruction, confronting the diverse facets of that disarray from different angles.

Part II. Complement and Supplements

As Susanna Lindberg reminds us in her chapter in this volume, "On Prosthetic Existence: What Differentiates Deconstruction from Transhumanism and Posthumanism", Jacques Derrida would draw our attention to "le manque", the want, the deficiency, proper to us and to any creature as there is indeed a constantly recreated shortcoming in us all calling for the replenishing of depleted resources: very soon we'll need to breathe again, sometime later in the day we will need to drink and to eat. All life needs to draw energy from the environment and in particular, breathe and feed itself. Our nature is in a constantly renewed need for rebuilding an ever-elusive completion.

Our genius lies in our successes at fixing or establishing at least on a permanent basis a partial relief of such recurrent reminders of our essential incompletion.

We gear to our surroundings through constant invention of new *supplements*, in attempts at remedying once and for all our inherent and ever-returning deficiencies. The squirrel deals with his by creating troves of acorns or hazelnuts, and in the same way, we discovered how to build granaries.

According to Rousseau, language arose as one of those *supplements*, as a manifestation of human *perfectibility,* then writing as a *supplement* to language, an observation that Derrida would revive and make contemporary again emphasising that, in addition to being a compensatory response to a lack, the *supplement* is as well a representative that deputises or supplants the original.

Any type of tools, once designed and manufactured, would constitute a new supplement to who we are. Over the ages humankind kept adding similar new layers of *supplements*. A *supplement* is however no *complement* as we could have done without. It is a key part of the genius of a species to have come up with a constant flow of *supplements*. That, in a way similar to Rousseau's to whom, after having thought he was possibly about to die, every new day came along as a welcome addition, a wonderful *supplement*. His life had by then been fulfilled: it was full already, without any need therefore for further *complements*; anything that would come in addition would only be, and could only be therefore, not only a welcome but also a wondrous *supplement*.

But such a distinction is not as clear-cut as one might imagine: the robin is in need of berries, of beetles, of worms, of caterpillars to eat; he is on the prowl, looking for ways to complement himself (it unfolds every day and he is hungry again). But if he uses a pine needle to dislodge a caterpillar from a bark's crevice, he has responded adequately to one of his wants having resorted to a technical trick: he has turned to a *supplement*. Yet as far as he's concerned, he has not noticed any difference between this particular method for finding food and any other that he usually resorts to.

Should a robin share with us the notion of a personal identity, at no point would he consider when grabbing a pine needle that he is trespassing the boundary of that personal identity in the way we human beings do when having performed a particular heroic deed, made a revolutionary invention or proposed a new paradigm. When in *A World on the Wane* (1955), Claude Lévi-Strauss recalls his writing of a play entitled *The Apotheosis of Augustus*, he describes aptly such a human experience of feeling that one has acquired, if only for a fleeting moment, the status of a semi-god.

There are, therefore, two possible images of human nature: one where it is incomplete and in constant need to be completed because of its shortcomings, and another, where Man has kept adding new layers, which are as many *supplements*. Looking at the global picture, one realises that it is constantly enriched in that respect.

Part III. Boundaries and Frontiers

Posthumanism is militantly anti-elitist, so much so that it extends the compass of universalism itself. It is this time the "dominion over the fish of the sea, and over the fowl of the air, and over the cattle, and over all the earth, and over every creeping thing that creepeth upon the earth" that gets blown up. The realm of dignity reserved so far to *Homo Imago Dei* needs now to be shared with other creatures or entities. Stefan Lorenz Sorgner in the debate we hold in the present volume suggests an extension of dignity to a wide-ranging swathe of creatures, retaining the capacity to suffer as a valid qualifying criterion, while at the same time defending the view that the notion of truth is vacuous. However, when choosing as an example of a debate around dignity, Sorgner turns to the case of an orangutan whom a court in Argentina set free. But how could any justice system, such as the Argentinian appeal court in question, operate without truth and falsehood constituting its frame of reference? If we think of the concept of responsibility whose three strands are imputability, accountability and answerability, imputability assumes that the culprit is truly the cause of the unfortunate course of events which justify his or her presence in court, accountability supposes that the accused or witness has the capacity to provide a true account of what happened, while answerability entails that the witness or accused in the box has the capacity of distinguishing truly between Good and Evil.

However generous the ambitions of posthumanism at expanding dignity beyond the borders of the human species, it needs to be noted that large portions of us humans do not even extend the notion of dignity to humankind as a whole, reserving it to members of one's own ethnicity, to people sharing one's own creed, or confining it to even smaller groups such as a band of kin-related folks. The somewhat nowadays paradoxical situation is that while posthumanism aims to extend dignity further than the humanistic *Homo Imago Dei* starting point, the current rise of nationalistic or regionalist populisms sees the extent of the "Us" receding in most quarters of the globe. It needs to be added that as Roden (2020) convincingly recalls, not all posthumanisms have an ethical orientation—*speculative posthumanism* for one explicitly recuses from the ethical engagement of critical posthumanism.

In her contribution to this volume, "When Skin and Technology Intertwine", Hélène Jeannin raises the question of boundaries under a different guise: the traditional barrier of the skin is getting porous as humans are ever more ready to acknowledge the power of piercing and breaking one's skin to other professionals (or amateurs) than surgeons who were once the only ones being assigned that powerful invasive right. There is indeed a trivialisation of prostheses' status for purposes which would have been regarded as frivolous until recently, such as implanting a sub-cutaneous chip in order to easily open a door. A reflection of this type is reminiscent of Lucien Lévy-Bruhl's observation in the 1920s that for the so-called primitive mentality (in truth, the cultural sphere of archaic China) the notion that the person is fully contained within the skin's boundary is anathema: personal identity extends to a wide range of "appartenances" (*belongings*), and more specifically to one's *totem* and every one of its worldly and cosmological incarnations (Jorion 2011: 293–295).

Part IV. The Enlightenment Recovered

Transhumanism is a reaction to the second crisis of humanism, that linked to the questioning of the Enlightenment's blind faith in progress. The destructive power of the atomic bomb played a crucial role in that crisis. In the two-phased process summed up as "disarray and reconstruction", transhumanism has adopted the radical view of ignoring what had caused the disarray in the first instance so as to reconnect at the very location where the disconnect had taken place. Lindberg writes in the present volume: "instead of contesting traditional Enlightenment humanism, transhumanism actually adopts and enforces it". Would it be too daring to call therefore transhumanism "the Enlightenment recovered"?

Here lies besides the added value of Salomé Bour's chapter in this volume, "On Max More's Extropianism", first drawing our attention to the role played by Max More's "extropianism" as having broken the ground and provided the blueprint of later transhumanism and then, second, displaying to what large extent his manifesto promoted a version of libertarianism updated and boosted through having been rendered dynamic. Bour underlines that More was quick to soften somewhat his message by shifting his main references from the very radical sources of ultra-individualism that Ayn Rand and Murray Rothbard instantiated, to milder versions of it with Karl Popper and William Warren Bartley.

In a similar way that Bour emphasises the close-knit relationship of transhumanism with libertarianism as far as political thought is concerned, Alexander Thomas, in his contribution here, titled "'Transhumanism and Advanced Capitalism: Elitist Logics and Dangerous Implications", stresses the close-knit relationship existing between transhumanism and capitalism. Thomas is especially vocal when rebuking the ultra-individualistic premise that social and economic ills can be remedied merely through individual reform, doing away so with the differences existing between what can be changed through individual action and what implies modifying entire social structures, which can only be achieved through collective action.

Thomas writes from that standpoint about Julian Savulescu's way of thinking: "Savulescu simplifies all social inequality as being caused by human dispositions [and] puts the blame squarely within the moral core of each and every individual, all of whom need fixing. He offers little recognition of the social contingency of human moral failings". Thomas characterises Steve Fuller and Veronika Lipinska's approach thus: "They wish for a kind of permanent shock doctrine, disaster capitalism on acid, whereby 'the prospect of ecological collapse, epidemics or even global financial meltdown [might] serve a similar function to focus minds in our own day'".

In Thomas' opinion, the expulsions and concentrations we are seeing at work in advanced capitalism mean a small sub-population only plays a role in re-defining "Man".

Ultra-individualism displays an especially voluntarist view of humankind where human beings, in a first step make decisions and very unproblematically implement them in the real world in a second step, with supposedly very little wastage along the road. In my own contribution to this volume, entitled "Posthumanism, Transhumanism, Superhumanism and Metahumanism from an Adaptive Standpoint", I underline that those four varieties of neo-humanism are in essence all pre-Freudian in that particular respect of "voluntarism", relying each on an antiquated simple-minded "intention/decision/implementation" operational model of the human psyche and willpower.

That peculiarity of a pre-Freudian framework in all four current brands of neo-humanism is worth mentioning as every one of them owes much to the legacy of Nietzsche, the towering precursor of Freud, who sketched very much in all its aspects the ground-breaking discoveries assigned today to the inventor of psychoanalysis—if one excepts that is, Nietzsche's own historical inspiration: Paul of Tarsus, the arch-precursor who stressed unrelentingly the conflicts between the "soul" and the "flesh" or as we would express it in contemporary parlance: the *conscious* and the *unconscious*. In Joachim Köhler's uncompromising terms indeed: "Almost everything that is today associated with the name of Sigmund Freud can be found in the writings of Friedrich Nietzsche, if only one knows where to look" (2002 [1989]: 209). Köhler proceeds then by listing convincingly: the *Superego*, present with Nietzsche as "ancestors continuing to exist as powerful spirits", the "wave effect" for adult life of a trauma in childhood, the interference of the unconscious with the conscious mind: "The decisive value of an act lies precisely in its non-intentional quality" and repression: "Forgetfulness is no mere passive quality, the product of lassitude, but a positive power of holding back, of inhibiting, a guarantor of the preservation of spiritual order" (ibid.).

Part V. *The Cunning of Reason*

When invoking the *Cunning of Reason,* Hegel refers to these numerous occasions when the human race has made a decisive turn in its history despite no individual human holding at the time a clear representation of the process actually at work. Starting from the premise that there is active in every one of us a *Cunning of Reason* that the unconscious operates, I suggest in my own contribution here that there is currently effective a *Cunning of Reason* for the species as a whole for which transhumanism offers a discourse justifying the often very daring steps required to ensure its survival, whether that would occur on planet Earth or anywhere else in the universe.

In that same perspective, Clément Vidal and Francis Heylighen, in their contribution titled "Ethics and Complexity", stress the crucial role played by ethics in the survival of our species: "ethics is not about preserving the self, but about preserving the group", and in a perspective akin to that of posthumanism where "dignity" needs to be extended further away rather than just restricted to humans, they suggest extending the range of ethics altogether: "in our transforming and complexifying society [...] ethics needs its Copernican revolution to be able to deal with all moral agents, not only humans". A view which ties in neatly with Émile Durkheim's concept of ethics amounting to the "social intériorisé" (*the internalised social*) within each of us, unformulated precepts presiding to a way of life that goes without saying. Again, a notion overlapping with this inner instance which Freud attempted to capture with that of a *Superego*, an unconscious supervisor at work in us and making sure unrelentingly, often in a persecutory way, that a person, far from being one in a collection of so many Robinson Crusoes, is, in Aristotle's words, a *zoon politikon*: a social animal.

References

Canguilhem, Georges, *The Normal and the Pathological* [1943], Berlin: Springer, 1978

del Val, J., and S. L. Sorgner, *A* Metahumanist *Manifesto*, 2010. http://metabody.eu/wp-content/uploads/2016/02/A-METAHUMANIST-MANIFESTO.pdf

Foucault, Michel, *The Order of Things: An Archaeology of the Human Sciences* [1966], New York: Pantheon Books, 1970

Hugo, Victor, *Choses vues 1847–1848*, Paris: Gallimard, 1972

Jorion, Paul, *Le capitalisme à l'agonie*, Paris: Fayard, 2011

Köhler, Joachim, *Zarathustra's Secret. The Interior Life of Friedrich Nietzsche* [1989], New Haven, CT: Yale University Press, 2002

Lévi-Strauss, Claude, *Du miel aux cendres. Mythologiques II*, Paris: Plon, 1966

Plato, *Protagoras*, Indianapolis-New York: Bobbs-Merrill, 1956

Roden, David, 'Posthumanism: Critical, Speculative, Biomorphic', in Jacob Wamburg and Mads Thomsen (eds.), *The Bloomsbury Handbook of Posthumanism*, London: Bloomsbury, 2020

Rousseau, Jean-Jacques, *Œuvres complètes III*, Bibliothèque de la Pléiade, Paris: Gallimard, 1964

Rousseau, Jean-Jacques, *Essai sur l'origine des Langues*, texte intégral reproduit d'après l'édition A. Belin de 1817, Bibliothèque du Graphe (no indication of date or place), n.d.

Part I

Humanism on the Wane

Strong Artificial Intelligence and Theological Anthropology: One Problem, Two Solutions

Marius Dorobantu

A survey of the transhumanist narratives for the future of humanity reveals that one of the most important variables, alongside human enhancement, is artificial intelligence (AI). The extent to which AI can progress remains one of the big unknowns, and contemporary opinions range from scepticism to unrestricted optimism. However, recent developments and achievements have reopened the possibility of AI reaching human level in the not-so-distant future. This is a scenario so radical that it would challenge the most fundamental principles of our society and our understanding of the world.

In this chapter, the scenario of human-level AI is approached from a theological perspective. This is done by reviewing the questions it would pose to our understanding of the human being and of human uniqueness, which are theologically articulated in the concept of being created in the image of God (Latin, *imago Dei*).

Since the beginnings of the field in the 1950s, with the Dartmouth workshop and the very coining of the term "artificial intelligence" by

M. Dorobantu (✉)
Vrije Universiteit Amsterdam, Amsterdam, The Netherlands

P. Jorion (ed.), *Humanism and its Discontents*,
https://doi.org/10.1007/978-3-030-67004-7_2

John McCarthy, AI has had various ups and downs. The expectations fluctuated from unrealistically optimistic—with one of the AI pioneers supposedly being confident that the problem of computer vision could be solved in one summer[1]—to the very pessimistic moods during the several "AI winters," when *true AI* seemed to be at times forever out of reach.

Nowadays, we seem to be yet again in the middle of another "AI summer." After the turn of the twenty-first century, technological progress in hardware building has started to produce unprecedented powerful supercomputers. These, in turn, have finally enabled the functioning of so-called machine learning algorithms, based on deep artificial neural networks. The principles of these algorithms had already been figured out in the 1980s, but they had lacked the hardware support that would render them effective.

Some of the most impressive applications of AI today are in the fields of game-playing, image and voice recognition, language translation, or car navigation. However, impressive as they may be, they all still fall in the category of narrow, or so-called weak AI: they can do one type of task, sometimes even at super-human levels, but are unable to learn and execute something completely different.

Human-level AI, sometimes referred to as "strong AI," on the other hand, would have a high degree of generality, just as human intelligence does. It could apply the knowledge acquired in one field to a totally different field. It is believed that, with enough technological progress, strong AI could eventually perfectly master human behaviour, becoming virtually indistinguishable from us from the point of view of what it could do. This is, in fact, how the famous Turing Test is designed: when a program will be able to trick human judges to think they are chatting with a human being, then it will be considered true AI.[2]

There is a lot of debate whether that would mean that strong AI would have consciousness or a mind, whether it would be a person or not, with all the philosophical, ethical and legal consequences entailed by the notion of personhood. But these are not questions addressed in this chapter. Instead, it might be useful to turn the discussion towards ourselves, humans, and ask, from a theological perspective, what would such a scenario mean for our understanding of human nature and human uniqueness.

The possible emergence of strong AI could open up the universe to wholly new spectres of intelligence, ranging from human-level AI to super-human-level AI (the same capabilities of human intelligence enhanced to superb speeds and performances), up to artificial super-intelligence—something with capacities we cannot even think of, functioning at unimaginable levels.[3]

It is not yet clear if and when strong AI could emerge, but many of the people involved in AI research and development think there is a good chance it could happen in the next few decades. A survey done by Müller and Bostrom, among experts working in the field of AI, reveals a median 50% chance of human-level AI occurring before 2040.[4]

From a theological point of view, the most important question raised by this scenario regards human uniqueness and specialness, which have always been unequivocally affirmed by all Christian theological anthropologies. The special status of humans has always been rooted in the belief that we are created in the image of God, a belief explicitly affirmed at the beginning of the book of Genesis and developed ever since. The possibility of AI acquiring all the features once thought to be uniquely human and surpassing human intelligence seems to shatter the concept of *imago Dei* and even theological anthropology altogether.

The question tackled further is whether Christian theology can still coexist with such a radical scenario. Could it find the internal resources to interpret coherently such a development while still maintaining its current claims?

Christian theology, in all its variations, has always claimed a special place for humans within the created world. Genesis proclaims that humans are created in the image of God,[5] which distinguishes them from all the other creatures. Defining what exactly means to be *imago Dei* was never an easy task, and theologians have struggled with the question from Patristic times up until today.

For the most part of our history, though, it was still pretty easy to see a big gap between us and everything else: we were the only ones with rationality and self-consciousness, capable of complex language, culture, technology and love. However, current scientific convergence is narrowing down this gap. A few examples of challenges posed by modern science to the notion of the *imago Dei* follow.

First, from a biological point of view, we are in a perfect continuation of evolution with all life on earth; we share most of our DNA (around 99%) with our evolutionary cousins, chimpanzees.[6] Therefore, from this point of view, there is no ontological break in the chain of evolution. Also, it is extremely unlikely that at a single point in time a soul or a specific cognitive capacity was bestowed upon humans.[7] Theologically speaking, the question is:

> In what ways is human nature different from the nature of nonhumans and, therefore, like the nature of God?[8]

Second, humans seem to be much more similar to non-human animals than previously thought in terms of behaviour. Neither emotions[9] nor even morality[10] seems to be confined to humans. Instead, they are shared with at least some of the animals. Moreover, we have increasingly more proof of animal self-consciousness, in various degrees.[11]

Third, another challenge comes from paleoanthropology: non-Sapiens hominids, like the Neanderthals or the Denisovans, seem to have had similar brain sizes and to have possessed most, if not all, the cognitive features that used to be thought of as unique to *Homo Sapiens*.[12] We can only join Gregory Peterson in asking the crucial question: 'were they too in the image of God?'[13]

Last, but not least, neuroscience and evolutionary psychology claim to be making significant advances in explaining feelings and even consciousness. There is an increasing pressure from science to admit that a materialistic understanding can account for everything we can see and measure regarding humans. This is not to say that the material reality is all there is. It is just that from a scientific point of view there is no more need to postulate the existence of the soul or the mystical character of human consciousness: humans seem to be just extremely complex algorithms of information processing.[14]

Theological anthropology is therefore "attacked" from all sides. Some still hope that consciousness or free-will will never be fully explained.

This is, in my opinion, just another kind of "theology of the gaps," where theology finds refuge in whatever science cannot yet fully understand. Given all the above, the emergence of strong AI would, therefore, not necessarily pose new questions. It would, however, enforce the validity of the existing ones by forcing anthropology to stare in the face of a much more brutal and unequivocal reality. Higher intelligence may not necessarily trigger conscience and free-will. We could have amazingly intelligent artefacts that would still lack intentionality and a mind, and indeed this seems to be where AI development has led so far. But this would just be the easy answer, preserving human uniqueness in the realm of whatever the machines do not yet possess.

But what if strong AI could acquire conscience, volition or any other feature that is generally part of a definition of personhood? This hypothesis may not be that far-fetched. After all, we currently have only one data point on the Cartesian graph of high intelligence and consciousness, that of homo sapiens. In our case, advanced intelligence does indeed coincide with all the other features mentioned above. If consciousness and personhood have emerged in connection with higher intelligence in the case of humans, why would not the same thing occur during the evolution of intelligent machines? If this happens, *the* question then becomes: what would this mean for how we understand human nature and human distinctiveness?

The implications for Christian theology are huge. The ideas of human uniqueness and the *imago Dei* sit at the core of our anthropology and are interrelated with basically everything else: they are embedded not only in protology, but also in Christology and eschatology. In fact, it is difficult to find any branch of theology that is not connected with the notion of the divine image. Could theology find a way to accommodate the implications of such a possibility as human-level AI? Or will the entire theological edifice just crumble? The question can thus be rephrased as follows: is it still possible to speak of human uniqueness if machines acquire the same qualities we have always regarded as the *imago Dei* in us? Two solutions to this problem are further explored, one more radical, and the other moderate.

The Radical Solution: Partially Abandon the Emphasis on the Concept of the *Imago Dei*

One handy solution could be to lower the emphasis assigned to the concept of *imago Dei* in our theological anthropology. This proposal could find some strong exegetical support.

First, the image of God occurs in the Old Testament only a few times, and all of the occurrences are in the book of Genesis. A legitimate question would be whether our theology has not made too much out of it. David Fergusson argues that it was precisely its placement at the beginning of the Hebrew Bible that made Christian theologians attribute it a disproportionate attention.[15] Moreover, there is a good exegetical case that the text should not even be at the beginning of the Bible in the first place. Graeme Auld argues that Genesis 1 is a later prologue to a book that used to start with Genesis 5:1.[16] If this is indeed the case, then it would reinforce the view that the concept of the *imago Dei* has been wrongly overemphasised because of its position.

Second, even if the presence of the *imago Dei* at the beginning of the Bible is legitimate, our use of it as a main source for theological anthropology could be mistaken. Joshua Farris points to the fact that the nature of the *imago* is incidental to the direct teachings found in the text.[17] Alistair McFadyen doubles down on this, arguing that our approach of regarding Genesis 1–3 as a 'deposit of answers to the anthropological question' is misguided from the start.[18] Since the account in Genesis is primarily interested in God and God's presence, not in humans, we should be careful when trying to use it as a source for anthropology.

The brief conclusion from these exegetical arguments is that the notion of *imago Dei* might be a little bit overemphasised in our theological anthropology, compared to its biblical support, which is rather weak.

The advantages of this so-called radical solution are obvious: by removing the *imago Dei* from the focal point of its anthropology, Christian theology could better accommodate not only scenarios of strong AI (and even extra-terrestrial life), but also the above-mentioned challenges from evolutionary science. It would mean that humans are not necessarily distinctive in a strong sense and, theologically speaking, there is no real problem with this. Moreover, by removing the burden of uniqueness, this acceptance could enable humans to think more solidarily about the rest of creation, and it could liberate them to envisage their destiny more creatively.

However, this would mean a huge leap for Christian theology, with powerful implications for all its branches, which would need to be rebuilt almost from scratch. *Imago Dei* has been so central of an issue that it seems very difficult to even think of an adequate theological anthropology without it. Besides, this proposal is susceptible to the criticism of surrendering too easily something that is too important. Once down on this pathway, theology could end up giving up too much, until the point where it became irrelevant, dissolving in the post-modern landscape.

The Moderate Solution: Settle for a Non-ontological Interpretation

A more attentive analysis of the dialogue between theology and science on the topic of *imago Dei* reveals that the main problems do not come from the concept itself, but from its ontological, or substantive, interpretation. This interpretation states that the image of God is a certain quality or set of capacities that reside within humans.

One of the problems is that science is closing down on what have so far been considered the mysteries of life, conscience and feelings, "decoding" them as just a different kind of algorithms for information processing. Another problem is that the ontological interpretation has 'a tendency to view some human beings as more exemplary of the divine nature than others,'[19] which is ethically wrong. If we only think about persons with severe cognitive disabilities, we may ask a difficult, but legitimate, question: are they not in the image of God just because they are not intelligent or capable of making decisions?

Besides the ontological interpretation, textbook theological anthropology[20] also speaks of two other big alternatives: the functional and the relational interpretations. The functional interpretation means that the *imago Dei* consists of our election by God to represent God in creation, by exercising dominion and stewardship over nature. God appoints humans as rulers over the created world. This view is supported by historical-critical biblical exegesis: in the historical context of the Ancient Near East, the king was seen as the image of god, meaning that he was God's appointed representative, having the right to act on God's behalf.[21]

The relational interpretation locates the image of God in the relations we establish and maintain with God and with each other. It is most notably proposed by Karl Barth. According to him, the fact that humans are created in relation—with God and with one another—constitutes the essence of the divine image, because it mirrors the very nature of God, the Trinity, as relation.[22] The image is thus in the relationship itself, not in our capacity for relationship.[23]

These two interpretations, although eliciting some legitimate criticism, seem indeed to be better equipped to answer the questions raised in the current dialogue between theological anthropology and evolutionary science. However, a scenario of strong AI would be a completely different beast, and a case can be made that neither of them could offer a satisfying account of human uniqueness in such a case.

The functional interpretation speaks of our election by God and of our mission of dominion and stewardship over creation. But why would intelligent robots not fulfil the same role, maybe even better than humans do? In describing how an artificial super-intelligence would function, philosopher Nick Bostrom mentions, among others, its role as a super-efficient sovereign of the world.[24] With access to the right resources, tools and decision mechanisms, a super-intelligence could govern the world much more efficiently than humans currently do. Not only would it preside over a more efficient world economy, but it would also end notorious big problems, such as poverty, disease or climate change. It would not only give us better lives, but it would also take care of the animals and of the planet in general. This seems to make it fit very well the functional definition of the *imago Dei*, certainly better than humans. This might sound far-fetched and naively optimistic, and maybe rightly so, but the scenario has to be considered, at least in principle.

The relational interpretation, on the other hand, roots our uniqueness in our capacity to be in relationship with God and with one another, in our status as a counterpart to God, a "you" to whom God can address and who can respond back. But if AI becomes strong, by the very definition it would outsmart humans in all the intellectual tasks, including dialogue, relationships, and emotional and social intelligence. The very essence of the Turing Test is relational: a machine should trick a human through its relational abilities. The question of whether the relationships developed by strong AI—with God, with humans and with other AIs—would truly be authentic deserves further exploration. But at first glance, there is no obvious reason why this possibility should be discarded.

Since neither a substantive interpretation nor its most established non-ontological counterparts seem to be able to satisfyingly accommodate the scenario of strong AI, we are forced to dig even deeper, to an interpretation that is often relegated to the category of "others," namely the eschatological interpretation. It affirms that the ideal relationship of human beings to God should not be located in the paradisiac past, nor in the present, but instead in the eschatological future.

An illustrious representative of this stream of thought is Wolfhart Pannenberg.[25] He sees the image of God related to *exocentricity*, an intrinsic disposition of human nature that points humans towards a destiny that has not yet been reached.[26] This exocentricity is what drives humanity towards its final becoming an image of God and towards its future destiny of fulfilling this thirst for transcendence in fellowship with God.[27]

Exocentricity might sound related to self-transcendence, another of the usual candidates for conveying human distinctiveness, as argued by Reinhold Niebuhr.[28] However, as argued by J. Wentzel Van Huyssteen, Pannenberg's exocentricity goes far beyond Niebuhr's notion of self-transcendence:

> Exocentricity not only points to a human orientation to others and to the world, but in a much more holistic sense reveals a disposition of human nature itself. Humans have this *Weltoffenheit,* or openness to the world, which by far transcends the openness of all animals to their environment. (…) Exocentricity thus means that humans are always open beyond every experience and beyond any given situation, in fact beyond the world itself. We are even open beyond our own cultural constructions: as we transform nature into culture, and constantly replace earlier forms of culture with new ones, we are also open beyond culture to the future, and to our finding our ultimate destiny in the future. This restlessness of human nature forms an important root for all religious life.[29]

What this means is that we are destined to always reach beyond everything we experience in the world in our search for fulfilment. *Imago Dei* is our source of direction in our quest towards our destiny, and it is already manifested theologically in the person of Jesus Christ.[30] *Imago Dei* is thus both a gift already given and a future destiny for humankind.

A complementary approach to Pannenberg's exocentricity, in the same realm of eschatological interpretations, is Ted Peters' *imago Dei* as the end of evolution.[31] Peters too chooses to add the crucial Christological layer to a correct understanding of *imago Dei*.

He brings together arguments from theology and science to prove that a good definition of how humans are is conditioned more by the future than by the past.[32] Theologically, human nature is not fixed at creation, but it allows for evolution and 'growth in Christlikeness': 'We live now as the *imago Dei* insofar as we live in him, insofar as we participate in the reality of the eschatological resurrection into the new creation.'[33]

Further, he draws upon arguments linked to theodicy: the current state of the world, with so much suffering and premature death, cannot be the final version of creation by a good God. He agrees with John Haught that cosmic and biological evolution point to the conclusion that our universe 'is in great measure not yet created.'[34]

Finally, Peters' interpretation of the *imago Dei* echoes Pannenberg's in its emphasis on vocation, or the 'divine call forward,'[35] which locates the image of God in the future, rather than in the past, with the eschatological Christ as its source.

The eschatological interpretation looks like a much better candidate to respond to the challenges of a strong AI scenario. One of its advantages is that it does not locate the image of God in a capacity that humans possess, as the ontological interpretation, or in something that they are able to do, as the functional and relational interpretations. From this point of view, we can very easily imagine strong AI that would outsmart humans all across the board, and still lack the exocentricity and vocation to fulfilment in Christ.

But, maybe even more importantly, it represents a proposal that is at the same time both profoundly theological and very well-articulated scientifically. Especially in Peters' model, we can notice how the existential restlessness and gravitational attraction towards the eschatological Christ do not need to be supernatural additions to human nature. They are part and parcel of our constitution through the specific evolutionary history of our species, which uniquely combines biological and cultural factors. First, this can facilitate the interdisciplinary dialogue between theology and science on the topic of human uniqueness. Second, if the *imago Dei* is deeply eschatological and specific to our unique type of biological evolution, it means that artificial beings would most likely fall outside its scope, unless they undergo a similar evolutionary trajectory, which might be astronomically unlikely. In other words, they will probably not be in the image of God, regardless of their high level of intelligence.

In conclusion, reflecting right now upon the scenarios of strong AI is certainly important for theology and it can force us to re-interpret our tradition in a fresh way, exploring the truth of divine revelation from new, original angles. Theologian Aubrey Moore famously said about the relation between Darwinism and Christianity: 'Darwinism appeared, and, under the guise of a foe, did the work of a friend.'[36] The same could be true of strong artificial intelligence.

Acknowledgements The research for this paper was made possible through the support of a grant from the Templeton World Charity Foundation. The opinions expressed in this publication are those of the author and do not necessarily reflect the views of the Templeton World Charity Foundation.

Notes

1. Seymour Papert, "The Summer Vision Project," *MIT AI Memos (1959–2004)*, 1966. hdl:1721.1/6125. Available at https://dspace.mit.edu/handle/1721.1/6125 (Retrieved June 21, 2020).
2. Alan Turing, "Computing Machinery and Intelligence," *Mind* 49: 433–460, 1950.
3. Nick Bostrom, *Superintelligence: Paths, Dangers, Strategies*, Oxford, UK: Oxford University Press, 2014.
4. Vincent C. Müller and Nick Bostrom, "Future Progress in Artificial Intelligence: A Survey of Expert Opinion," in *Fundamental Issues of Artificial Intelligence*, edited by Vincent C. Müller, Berlin, DE: Springer, 553–571.

5. Gen. 1: 27.
6. Robert H. Waterson, Eric S. Lander, and Richard K. Wilson, "Initial Sequence of the Chimpanzee Genome and Comparison with the Human Genome," *Nature* 437: 69–87, 2005.
7. Aku Visala, "Imago Dei, Dualism and Evolution: A Philosophical Defense of the Structural Image of God," *Zygon: Journal of Religion and Science* 49 (1): 101–120, 2014, p. 114.
8. Colin E. Gunton, "Trinity, Ontology and Anthropology: Towards a Renewal of the Doctrine of the *Imago Dei*," in *Persons, Divine and Human*, edited by Christoph Schwöbel and Colin E. Gunton, Edinburgh, UK: T&T Clark., 1991, 47–61, p. 47.
9. Frans de Waal, *Mama's Last Hug: Animal Emotions and What They Tell Us about Ourselves*, W.W. Norton & Company, 2019.
10. Frans de Waal, *The Bonobo and the Atheist: In Search of Humanism Among the Primates*, W.W. Norton & Company, 2013.
11. Gordon G. Gallup Jr., "Chimpanzees: Self-Recognition," *Science* 167 (3914): 86–87, 1970; Diana Reiss and Lori Marino, "Mirror self-recognition in the bottlenose dolphin: A case of cognitive convergence," *Proceedings of the National Academy of Sciences* 98 (10): 5937–5942, 2001.
12. Sverker Johansson, "The thinking Neanderthals: What do we know about Neanderthal cognition?", *WIREs Cognitive Science* 5: 613–620, 2014.
13. Gregory R. Peterson, "Uniqueness, the Image of God, and the Problem of Method: Engaging Van Huyssteen," *Zygon: Journal of Religion and Science* 43 (2): 467–474, 2008, p. 473.
14. Steven Pinker, *The Blank Slate: The Modern Denial of Human Nature*, Penguin, 2012.
15. David Fergusson, "Humans Created According to the Imago Dei: An Alternative Proposal," *Zygon: Journal of Religion and Science* 48 (2): 439–453, 2013, p. 442.
16. Graeme Auld, "Imago Dei in Genesis: Speaking in the Image of God," *The Expository Times* 116: 259–262, 2005.
17. Joshua R. Farris, "An Immaterial Substance View: Imago Dei in Creation and Redemption," *The Heythrop Journal* LVIII: 108–123, 2017, p. 108.
18. Alistair McFadyen, "Imaging God: A Theological Answer to the Anthropological Question?" *Zygon: Journal of Religion and Science* 47 (4): 918–933, 2012, p. 923.
19. Fergusson, "Humans Created According to the *Imago Dei*...," p. 443.
20. Marc Cortez, *Theological anthropology: a guide for the perplexed*, New York, NY: T&T Clark, 2010, pp. 14–40; Noreen Herzfeld, *In Our Image:*

Artificial Intelligence and the Human Spirit. Minneapolis, MN: Fortress Press, 2002, pp. 10–30.

21. John H. Walton, "Human Origins and the Bible," *Zygon: Journal of Religion and Science* 47 (4): 875–889, 2012.
22. Karl Barth, *Church Dogmatics* Vol. 3, Edinburgh, UK: T&T Clark, 1958, pp. 184–185.
23. Herzfeld, *In Our Image…*, p. 26.
24. Bostrom, *Superintelligence…*, pp. 177–193.
25. Wolfhart Pannenberg, *Anthropology in Theological Perspective*, Philadelphia, PA: T&T Clark, 1985, pp. 43–79.
26. LeRon F. Shults, *Reforming Theological Anthropology: After the Philosophical Turn in Relationality*, Grand Rapids, MI: Eerdmans, 2003, p. 235.
27. J. Wentzel van Huyssteen, *Alone in the World? Human Uniqueness in Science and Theology*, Göttingen, DE: Vandenhoeck & Ruprecht, 2006, p. 139
28. Reinhold Niebuhr, *The Nature and Destiny of Man: A Christian Interpretation*, Louisville, KY: Westminster John Knox, 1966, vol. 1, pp. 161–162.
29. Van Huyssteen, *Alone in the World…*, pp. 139–140.
30. Ibid., p. 141
31. Ted Peters, "The *Imago Dei* as the End of Evolution," in Stanley P. Rosenberg, Michael Burdett and Benno Van Den Toren (eds.), *Finding Ourselves After Darwin*, Grand Rapids, MI: Baker Academic, 2018, 92–106.
32. Ibid., p. 94.
33. Ibid., p. 95.
34. John Haught, *Deeper than Darwin*, Boulder, CO: Westview, 2003, p. 168.
35. Peters, "The *Imago Dei…*," p. 96
36. Francisco J. Ayala, *Darwin's Gift to Science and Religion*, Washington, DC: Joseph Henry Press, 2007, p. 159.

References

Auld, Graeme (2015). 'Imago Dei in Genesis: Speaking in the Image of God.' *The Expository Times* 116: 259–262. https://doi.org/10.1177/0014524605052835.

Ayala, Francisco J. (2007). *Darwin's Gift to Science and Religion*. Washington, DC: Joseph Henry Press.

Barth, Karl (1958). *Church Dogmatics*. Vol. 3. Edinburgh: T&T Clark.
Bostrom, Nick (2014). *Superintelligence: Paths, Dangers, Strategies*. Oxford: Oxford University Press.
Cortez, Mark (2010). *Theological Anthropology: A Guide for the Perplexed*. New York, NY: T&T Clark.
de Waal, Frans (2013) *The Bonobo and the Atheist: In Search of Humanism Among the Primates*. W.W. Norton & Company.
de Waal, Frans (2019). *Mama's Last Hug: Animal Emotions and What They Tell Us about Ourselves*. W.W. Norton & Company.
Farris, Joshua R. (2017). 'An Immaterial Substance View: Imago Dei in Creation and Redemption.' *The Heythrop Journal* LVIII, 108–123. https://doi.org/10.1111/heyj.12274.
Fergusson, David (2013). 'Humans Created According to the Imago Dei: An Alternative Proposal.' *Zygon: Journal of Religion and Science* 48 (2): 439–453. https://doi.org/10.1111/zygo.12014.
Gallup, Gordon G. Jr. (1970). 'Chimpanzees: Self-recognition.' *Science* 167 (3014): 86–87. https://doi.org/10.1126/science.167.3914.86.
Gunton, Colin E. (1991). 'Trinity, Ontology and Anthropology: Towards a Renewal of the Doctrine of the *Imago Dei*.' In *Persons, Divine and Human*, edited by Christoph Schwöbel and Colin E. Gunton. Edinburgh, UK: T&T Clark. 47–61.
Haught, John (2003). *Deeper Than Darwin*. Boulder, CO: Westview.
Herzfeld, Noreen (2002). *In Our Image: Artificial Intelligence and the Human Spirit*. Minneapolis, MN: Fortress Press.
Johansson, Sverker (2014). 'The Thinking Neanderthals: What Do We Know about Neanderthal Cognition?' *WIREs Cognitive Science* 5: 613–620. https://doi.org/10.1002/wcs.1317.
McFadyen, Alistair (2012). 'Imaging God: A theological Answer to the Anthropological Question?' *Zygon: Journal of Religion and Science* 47 (4): 918–933. https://doi.org/10.1111/j.1467-9744.2012.01291.x.
Müller, Vincent C. and Nick Bostrom (2016). 'Future Progress in Artificial Intelligence: A Survey of Expert Opinion.' In *Fundamental Issues of Artificial Intelligence*, edited by Vincent C. Müller. Berlin: Springer. 553–571. https://doi.org/10.1007/978-3-319-26485-1_33.
Niebuhr, Reinhold (1966). *The Nature and Destiny of Man: A Christian Interpretation*. Vol. 1. Louisville, KY: Westminster John Knox.
Pannenberg, Wolfhart (1985). *Anthropology in Theological Perspective*. Philadelphia, PA: T&T Clark.

Papert, Seymour (1966). 'The Summer Vision Project.' *MIT AI Memos (1959–2004)*. hdl:1721.1/6125. Available at https://dspace.mit.edu/handle/1721.1/6125, consulted on June 21st, 2020.

Peters, Ted (2018). 'The *imago Dei* as the End of Evolution.' In *Finding Ourselves After Darwin*, edited by Stanley P. Rosenberg, Michael Burdett and Benno Van Den Toren. Grand Rapids, MI: Baker Academic. 92–106.

Peterson, Gregory R. (2008). 'Uniqueness, the Image of God, and the Problem of Method: Engaging van Huyssteen.' *Zygon: Journal of Religion and Science* 43 (2): 467–474. https://doi.org/10.1111/j.1467-9744.2008.00928.x.

Pinker, Steven (2012). *The Blank Slate: The Modern Denial of Human Nature*. Penguin.

Reiss, Diana and Lori Marino (2001), 'Mirror Self-recognition in the Bottlenose Dolphin: A Case of Cognitive Convergence.' *Proceedings of the National Academy of Sciences* 98 (10): 5937–5942. https://doi.org/10.1073/pnas.101086398.

Shults, LeRon F. (2003). *Reforming Theological Anthropology: After the Philosophical Turn in Relationality*. Grand Rapids, MI: Eerdmans.

Turing, Alan (1950). 'Computing Machinery and Intelligence,' *Mind* 49: 433–460.

van Huyssteen, J. Wentzel (2006). *Alone in the World? Human Uniqueness in Science and Theology*. Göttingen, DE: Vandenhoeck & Ruprecht.

Visala, Aku (2014). 'Imago Dei, Dualism and Evolution: A Philosophical Defense of the Structural Image of God.' *Zygon: Journal of Religion and Science* 49 (1): 101–120. https://doi.org/10.1111/zygo.12063.

Walton, John H. (2012). 'Human Origins and the Bible.' *Zygon: Journal of Religion and Science* 47 (4): 875–889. https://doi.org/10.1111/j.1467-9744.2012.01301.x.

Waterson, Robert H., Eric S. Lander, and Richard K. Wilson (2005). 'Initial Sequence of the Chimpanzee Genome and Comparison with the Human Genome.' *Nature* 437: 69–87. https://doi.org/10.1038/nature04072.

Part II

Complement and Supplements

On Prosthetic Existence: What Differentiates Deconstruction from Transhumanism and Posthumanism

Susanna Lindberg

One of the most obvious characteristics of our epoch is the omnipresence of a rapidly changing technological environment that is becoming increasingly intimate and ubiquitous. It becomes more and more intimate to our bodies, which are not only *equipped* by always new smart devices, but one could almost say *inhabited* by ever more discreet medical aides that function as internal prostheses (for instance, insulin pumps, organ transplants, electronic stimulators and pharmacological regulators located deep within the organism). In principle, with the gene scissors CrisprCat9, some elements of human bodies might soon be *produced* technologically even before the persons are actually born. On the other hand, the disembodied (not to say "spiritual") dimension of intellectual and social life is increasingly affected by the impersonal ubiquitousness of digitalisation that mimes and parasites thought and communication. While yesterday's science fiction thus becomes reality, today's science

S. Lindberg (✉)
University of Leiden, Leiden, The Netherlands

P. Jorion (ed.), *Humanism and its Discontents*,
https://doi.org/10.1007/978-3-030-67004-7_3

fiction imagines technological futures in which today's humanity itself gradually turns obsolete.

Can technological change become so significant that it actually puts an end to humanity and changes it into something else, into a condition that has sometimes been called "transhuman" or "posthuman"[1]? Can intense technological transformations finally amount to anthropological transformations—knowing that, after all, technics is used on human beings by the human beings themselves? In what follows, I will investigate the philosophical conditions of this question by asking *what is the role of technics in the self-constitution of the human being.*[2] I will examine the question through a philosophical discussion between, on the one hand, the intellectual currents called "transhumanism" and "posthumanism", that in principle affirm the possibility of overcoming humanity by technics and, on the other hand, the philosophical discourse of "poststructuralism" and more precisely of "deconstruction", that does not affirm it. In what follows, these "-isms" are used in a very schematic way only to indicate opposing positions. At the end, at stake is not the overcoming itself but the conceptions of humanity and technicity that the question mobilises.

Examining a debate presupposes identifying its participants. In the present case, this is quite difficult, since especially "transhumanism" and "posthumanism" are not fixed notions but historically changing positions that often evolve by absorbing their critiques. Some versions of transhumanism can be assimilated into posthumanism,[3] and Stefan Lorenz Sorgner actually suggests that they could be united into a "metahumanism" (Ranisch and Sorgner 2014, pp. 34, 301). This is why it has been possible to claim that both trans- and posthumanism continue postmodernist and poststructuralist ideas (Besnier 2012, p. 165; Ranisch and Sorgner 2014, pp. 40–43, 50). Sometimes, on the contrary, transhumanism is allied to analytical philosophy and opposed to poststructuralism (More in More and Vita-More 2013, p. 10), while many versions of posthumanism count poststructuralism and postmodernism as their predecessors. But as the terms poststructuralism and postmodernism are also quite nebulous, the rapprochement is true of certain authors (e.g. Braidotti 2013; Herbrechter 2014), while it is incorrect of others (as, for instance, Wolfe (2010) and Neyrat (2015) have shown).

In what follows, I will attempt to shed light on this hazy domain through the mirror of the question concerning the role of technics in the self-constitution of the human being. What interests me in the end is not the interlocutors but only this question that the debate hopefully allows me to clarify. In this chapter, I will first make a schematic distinction between trans- and posthumanist conceptions of technics as reflected in their conceptions of humanity. Then, instead of speaking about post-structuralism in general (there hardly is such a thing), I will present three French philosophers, Jacques Derrida, Jean-Luc Nancy and Bernard Stiegler, that have often been ranged under the rubrics poststructuralism or deconstruction, although if asked, they would reject such labels, albeit for different reasons. The article concentrates on them because through their works we can follow the emergence of an idea of human existence as originary technicity. They think technics as the "distinctive feature of humanity" (*propre de l'homme*) while at the same time denying the very possibility of defining anything like a "distinctive feature" (*propre*) or "humankind" (*l'homme, l'humain, l'humanité*). In different ways, they think technicity of existence and Stiegler even formulates a kind of a techno-anthropo-logy, but they do not call for a new "humanism" (not even, in these terms, for a trans- or posthumanism[4]). I will start by presenting Derrida's account of prosthetic existence, then show how it has been completed by Nancy's account of the eco-technical intruder, and conclude by explaining how this has finally been developed into a full-blown philosophical anthropology in Stiegler's theory of technics as epi-phylogenetic memory. In the end, we will see why this line of thoughts is incompatible with transhumanism and also with a certain understanding of posthumanism. My ultimate aim, however, is not this clarification (which remains instrumental) but the philosophical question of the role of technics in humanity (or ~~humanity~~).

Before entering my subject, still a word on method, for the following considerations suffer from a certain imbalance. I will speak about trans- and posthumanism on a quite general level as positions or discourses (cf. Herbrechter 2013, pp. 36–37). I believe this to be possible since the proponents of these views treat the terms "transhumanism" and "posthumanism" (or in Braidotti's case, "the posthuman") as central themes of their works, although there are of course significant differences on what these positions actually imply. My aim is not to make a comprehensive comparative study of different conceptions of transhumanism and posthumanism but only to identify, for the sake of the present argument, the hard core of these positions' understanding of the relation between technics and humanity. On the contrary, I cannot speak about deconstruction as a general discourse. Jacques Derrida, who coined the term, insisted that it is not a method and cannot be transformed into one (Derrida 2008): if a method shows a *possible* way of proceeding, deconstruction is the experience of the *impossible*, an event of undoing a text, the *happening* of a singular event of reading (Derrida 1992). Even Derrida did not call himself "deconstructionist", and his most obvious successors—here I will only name Jean-Luc Nancy and Bernard Stiegler, since these three understand human existence in terms of originary technicity—do not adopt the label. Derrida, Nancy and Stiegler have also sometimes been grouped under labels such as "poststructuralism" and "postmodernism" and also associated with "posthumanism", but in the case of these three, deconstruction is the least misleading of these equally misleading terms.

This is why the following considerations progress as a somewhat imbalanced comparison between general positions that assume the form of "isms" and texts that do not. The study of a philosophical text is methodologically easier because it can rely on the factuality of the text—even when it is a difficult one, like for instance Jacques Derrida's thinking of technics, and even when its reading leads to critical or deconstructive views on it. On the contrary, the study of a philosophical position is trickier. A position—like an "ism"—cannot be reduced into the factuality of a particular text. It is not a concept that frames an object of thought, either, but the manner in which the *subject* of a discourse is configured and the *ethos* that orients its specific ways of thinking, judging and acting. The position answers to the question of the subject of thinking, not in the ultimate philosophical way through enquiry into method, but in a more general way, by framing a specific configuration, including values, that orients the subject in both thought and life. It is much easier to analyse a text than to ("psycho")analyse the subjective position behind it, for the subject position is ultimately something that overflows the text. Hence, trans- and posthumanism are not univocal philosophical theses. They are general positions that are of course expressed in theses and arguments but that overflow them. So many changing and different discourses claim the nominations trans- and posthumanism that it is impossible to bring them into a univocal whole; instead, the reader must settle for their family resemblances, enlisting minimal common features and trying to seize the core *ethos* that holds them together.

Technics in Transhumanism and Posthumanism

Transhumanism is an intellectual position that advocates the radical transformation of the human being's biological, mental and social conditions by means of technology. The first article of the World Transhumanist Association's (WTA's) *Transhumanist Declaration* states, "Humanity stands to be profoundly affected by science and technology in the future. We envision the possibility of broadening human potential by overcoming ageing, cognitive shortcomings, involuntary suffering, and our confinement to planet Earth". In an earlier version of the *Transhumanist Declaration*, the WTA has also declared: "We support the development of access to new technologies that enable everyone to enjoy better minds, better bodies and better lives. In other words, we want people to be better than well" (Humanity + Website 2008 and 2010, quoted by Lilley 2013, p. 1). The transhuman philosopher Max More specifies that these "radical alterations in the nature and possibilities of our lives [result] from various sciences and technologies such as neuroscience and neuropharmacology, life extension, nanotechnology, artificial ultraintelligence, and space habitation" (Pilsch 2017, p. 1).

Transhumanists perceive such transformations as human *enhancement*, that is desirable as such, even if it was to amount to a form of life that is not human anymore. Robert Ranisch and Stefan Lorenz Sorgner explain, "The result of such technologically induced version of evolution is referred to as posthuman. However, there is no commonly shared conception of what posthumans are, and visions range from the posthuman as a new biological species, a cybernetic organism, or even a digital disembodied entity. The link between the human and the posthuman is the transhuman, an abbreviation for a transitional human, to which transhumanism owes its name" (Ranisch and Sorgner 2014, pp. 7–8, compare with More and Vita-More 2013, pp. 4, 28–53).

Although the posthuman may mean the result of such technological enhancement, the term also has another use, so that Cary Wolfe can say that it is the opposite of transhumanism (Wolfe 2010, p. xv). Posthumanism denotes a break with traditional humanism insofar as humanism is perceived as an ideological construct that paternalistically imposes Western "phallo-logo-centrism" over the whole world (and indeed, the transhumanist conception of "better minds, better bodies and better lives" incarnates very traditional Western ideas of good life). Many feminist, postcolonialist and ecologist authors call themselves posthumanists in order to oppose themselves to this sense of humanism (and to this ideal of mind, body and life). This is why the term has also been used to refer to authors who, like Donna Haraway, see the other of "man" to be a feminist figure of a technologically mediated "cyborg", as well as to others who, like Cary Wolfe, associate posthumanism rather to ecology than to technology or who, like Jane Bennett, is interested in the materiality of human and nonhuman life (Bennett 2010). To cut a long story short, one can situate transhumanism in Silicon Valley and posthumanism in departments of art and literature; or one can refer to transhumanism as a liberal-capitalist view on the becoming of humankind, and to posthumanism as a leftist one.

The origin of the term transhumanism is generally situated in the writings of Julian Huxley and of Pierre Teilhard de Chardin (Ranisch and Sorgner 2014, p. 9; Pilsch 2017, pp. 114–137). The modern sense of transhumanism was introduced in the mid-1960s by Fereidoun M. Esfandiary, who called himself FM-2030, and starting from the 1980s, the important theorists of transhumanism include Natasha Vita-More and Max More, Marvin Minsky, Eric Drexler, Hans Moravec and Ray Kurzweil. Today, transhumanism is defended by the World Transhumanist Association, founded by the philosophers Nick Bostrom and David Pearce in 1998, whose declarations are available on the Humanity+ website. Present-day transhumanists are a multifaceted group of engineers, scientists, futurist artists and utilitarian philosophers. They are bioliberals that think that biomedical technologies should be used in a non-therapeutic sense for the enhancement of human capacities such as health and longevity, emotional and cognitive capacities, physical traits and behaviour. Some of them dream about mind uploading and expect the advent of Singularity, which is the becoming-self-conscious of the infosphere (Kurzweil 2005).

The term posthumanism has several origins, as well. Its first use has been located in the Macy conferences on cybernetics (1946–1953). However, the term was properly coined by the postmodern philosopher Ihab Hassan in *Prometheus as Performer* (1977) (Tirosh-Samuelson in Ranisch and Sorgner 2014; Hayles 1999), who suggested that "human form may be changing radically... humanism may be coming to an end, as humanism transforms itself into something that we must helplessly call posthumanism". While this idea appears to be the same as the transhumanists' visions, Hassan's reference is different, namely the idea of the "end of man" that Michel Foucault formulated in 1966 (cf. Hagner and Hörl, 2008, p. 10; Neyrat 2015, p. 1). Foucault's famous sentence from *The Order of Things* goes as follows: "As the archaeology of our thought easily shows, man is an invention of recent date. And one perhaps nearing its end. If those arrangements were to disappear as they appeared [...] as the ground of Classical thought did, at the end of the eighteenth century, then one can certainly wager that man would be erased, like a face drawn in sand at the edge of the sea" (Foucault 2005, p. 422). Although this sentence certainly counts among the reasons of the confusion between trans- and posthumanism and poststructuralism, it actually does not have the same sense. Foucault does not mean to say that the biological species *homo sapiens* would change in any way, but that the idea of humanity is a historical construct that was invented as late as the eighteenth century and that might already be getting obsolete. In other words, Foucault's idea of the end of man is not the end of the human animal but of a particular self-image we hold of ourselves. The authors who first studied technological posthumanity, especially Donna Haraway and N. Katherine Hayles, agree with Foucault rather than with the transhumanists who take the transformation of the human species literally.

Now what is, according to these currents of thought, the role of technics in the self-constitution of the human being? The transhumanist position is quite clear. In most cases, transhumanist authors see technics as a means of human self-enhancement, and enhancement as a desirable aim as such. For instance, Max More affirms that "transhumanists regard human nature not as an end in itself [...] we can learn to reshape our nature in ways we deem desirable and valuable [...] we can become post-human. [...] Transhumanists refer to 'technology' as the primary means of affecting changes to human condition" (More and Vita-More, 2013, p. 4). In this definition, the essential point is the production of the human being by him/her/self: to explain it in Aristotle's familiar terms, in transhumanism, man sets the *final cause*, which is himself, as well as the *formal cause*, which is also himself, man is the agent of this production, and man himself is also the matter that is being formed. Technics is seen unproblematically only as the means of this production. Correlatively, the most well-known critical discussions of transhumanism have evaluated above all its aims. For instance, commentators and even works of science fiction have shown why the *final causes* of transhumanist production, such as immortality, absence of pain and ultra-rapid tireless brain, might actually not be desirable at all.[5] When one looks more closely into the transhumanist ideal of "better minds and better bodies", one realises that instead of contesting traditional Enlightenment humanism, transhumanism actually adopts and enforces it (More, in More and Vita-More, 2013, p. 10). Together with the enlightenment ideal, it also tends to reproduce its flip side, harsh productivism and hostility to incarnated existence. One can also doubt if the human being is so self-evidently the *agent* of transhumanist self-production. For who would that agent be? An individual might choose to improve himself in a certain manner, but in doing so, s/he can only choose among the technical dispositifs that are available in her/his society, which s/he cannot choose.[6] And finally, one can wonder if the human being can ever be regarded simply as the material (*hyle*) of the transhumanist production. Harnessing the body and the mind to preset aims might well destroy important parts of their own becoming in detrimental manners.

Such interrogations mainly arise from bioethical approaches to transhumanism. A very different, probably more primordial question arises when one interrogates technics itself instead of their human use. When technics is understood simply as a means, it is enough to evaluate the intentions of the human being that uses it. But what if technics produces effects of its own that cannot be reduced to any human intentions (like in the case of pollution)? Furthermore, what if it actually formats human intentions (like in the case of creation of new needs)? Martin Heidegger famously calls *Ge-stell* the fundamental structure of the modern technical world that disposes of man and nature before being at man's disposition (Heidegger 1977, pp. 24–25). Michel Foucault's term *dispositif* is a reinterpretation of *Ge-stell* especially in the context of social techniques. Both show how, not any particular technical system, but the general structure of the contemporary world, is a *technical framework* that predetermines what we can be and do and what we deem desirable. If technics is such a general frame, its effects can never be contained like the effects of simple means, because the very existence of means is already a consequence of the frame that conditions use and overflows it. The posthumanists who differentiate themselves from transhumanists by developing Foucault's (and Derrida's) idea of "the end of man" see technics in this way as a dispositif of a finite historical situation that conditions the human (and inhuman) beings that live in it. But at the same time, technics is also something that the human (and inhuman) beings can invent and use in different technics of self and of the world. Technics is really the endless mediation between existents and their situations.

One could maybe say that most versions of posthumanism are existential and political interpretations of this situation. Who are we, if we are inextricably bound to technics, formatted and formatting beings? What do we make of ourselves? For instance, Haraway's "cyborg" is a possible name for the existential situation of a being profoundly mixed with nonhuman life through endless technical mediations. In a parallel manner, Hayles locates posthumanity in the situation where human beings have given over one part of their thinking to technical systems and then identified themselves with the resulting hybrids. Does such posthumanity amount to overcoming humanity? At least it pays attention to the extension of humanity into domains that classical humanism rejected. But then again, what does "humanity" mean? In *A Thousand Plateaus*, Deleuze and Guattari note: "Why are there so many becomings of man, but no becoming-man? First because man is major-itarian par excellence, whereas becomings are minoritarian; all becoming is a becoming-minoritarian" (Deleuze and Guattari 1987, p. 313). This is why there is no becoming-man but only becoming-woman, becoming-child and becoming-animal. If humanism means only becoming-man in the majoritarian sense, then these posthumanist mixtures of human with nonhuman technicity represent the overcoming of humanity. But if the idea of the human being can accommodate its minoritarian becomings, as well, then the hybrids of hu/wo/man and machine reconfigure "humanity" instead of leaving it.

Technics in Deconstruction: Derrida, Nancy, Stiegler

Whatever the scope of "human", whatever continuity, critique or rupture with present humanity is indicated by the prefix "post", the literal meaning of the term "posthuman" points at a philosophical anthropology that, in a way or another, asks what is happening to the species/essence/figure "human". The most obvious difference between posthumanism and deconstruction is the fact that, although like all poststructuralists, philosophers practising deconstruction show how and why the old figure of humanity is disintegrating, they limit themselves to this deconstructive phase without opening the question of something (somebody, some figure of ~~human~~) that would come after, post, the human being. On the contrary, the word "posthuman" inevitably generates the idea of another species/essence/figure giving rise to another anthropology. The aim of deconstruction is not to propose another essence in the place of withered essences, but to work against the very idea of essentiality, at least when it comes to the self-interrogation of humans. Instead, deconstruction enquires into different ways of self-interrogation that are rooted in finite thinking. It examines the instance of discourse that is radically incommensurable with all anthropological essences and explores the finite instances of existence (me, you, other/s, us, etc.) that are eccentric in relation to all ideal types. The question of finitude keeps the thinkers of deconstruction apart from anything like a "human*ism*", even when the latter works in the direction of a sophisticated antiessentialism in order to define a post/trans/anti/in/humanism.

However, poststructuralism and deconstruction have certainly inspired posthumanism. One of the main reasons for this is their enquiry into originary technicity that has greatly contributed to the discovery of post/

humanity. Let us now see how Jacques Derrida, Jean-Luc Nancy and Bernard Stiegler have developed the motif of originary technicity.

Jacques Derrida's conception of technics seems to respond particularly well to the transhumanist question of man's technical enhancement and to the posthumanist question of human-technical hybrids, for he thinks technics primarily as *prosthetics* (for a detailed presentation, see Lindberg (2016)). After all, trans- and posthumanist bodies appear to be prosthetically enhanced bodies. Derrida shows that a prosthesis is not, or at least not simply, a means but—as we might say here—a dimension of subjectivation. By definition, a prosthesis patches a lack in a living body. It is an artificial addition to a living body, for instance, a walking stick or a pair of glasses, but also a pacemaker, a heart implant or a heart drug. A prosthesis is a piece of lifeless matter that is indispensable to life—a dead supplement added to a living body so that its living force would be enhanced, or quite simply so that the body could stay alive. Because the prosthesis is indispensable to life, it also reveals the weakness and the helplessness of bare life: if your heart was good, you wouldn't need a pacemaker, a heart implant or a heart drug, and more generally, if human life was self-sufficient, it would not need technics. In what follows, I will show how Derrida, Nancy and Stiegler interpret all kinds of technics in terms of prosthetics.

Derrida develops this theme ever since his early works, especially *De la grammatologie* from 1967 (Derrida 1976), in which his primary object is not technics in general, but one exemplary technique, namely *writing*. Like technics, writing should be understood here in the maximal sense including (1) the know-how of writing, (2) written objects, texts, and (3) entire systems, like the system of alphabetic writing or the technology of printing. Writing is particularly important for philosophy, because it is the technics that supplements and supports thought itself: it helps in the elaboration of thought, conserves thought from forgetfulness, and allows its sharing with other people. One could say that Derrida analyses writing as a prosthesis of thought in the sense that there is no thought that would not lean on language—that does not serve thought as a reliable tool, but haunts it as a prosthesis does. In this sense, one could also understand writing as a prosthesis of language and language as a prosthesis of thought. It should also be noted that writing can be analysed as a paradigmatic technique, but it is itself possible only on the ground of a more originary structure, *arche-writing*, that is not a technique, but the quasi-transcendental possibility of differentiation that conditions any system of writing as such. Arche-writing is Derrida's fundamental philosophical discovery, while here we are only interested in techniques of writing that are like phenomenal appearances of the non-appearing arche-writing. In *De la grammatologie*, Derrida also quotes the anthropologist André Leroi-Gourhan, who had recently brought about a new concept of technics, and makes an extensive reading of Jean-Jacques Rousseau. These will be studied in more detail by Derrida's student Bernard Stiegler in *La technique et le temps* (Stiegler 1998, 2009, 2010), but here I refer mainly to *De la grammatologie*, for the outline of this particular interpretation really comes from this book.

Derrida interprets *text* as an autonomous machine that is not a simple instrument of the author but has a functioning of its own parts from the same intuition as Gilbert Simondon's interpretation of the technical object. In his groundbreaking book *Du mode d'existence de l'objet technique* from 1958 (Simondon 2012), Simondon had shown that technics not only is a means of human intentions but has an irreducible mode of being of its own. It has its own logic and its own context that overflows its user's experience. Simondon defines the technical object as its capacity of functioning, which is dependent on its relation to other technical objects, technical elements and technical systems, and finally to its entire associated milieu, including humans. What the independence of the technical object means to the human being is shown by the paleo-anthropologist André Leroi-Gourhan, whom Derrida also comments in *De la Grammatologie*. Leroi-Gourhan thinks that human evolution is essentially technological and not biological or cultural (Leroi-Gourhan 1993). The technical object allows the conservation and the sharing of human thought: individuals can actually ignore the thought itself and yet be capable of using the object. For instance, one can know how to use a hammer without knowing how to fabricate one, not to speak about computers and cell phones. In *De la Grammatologie*, Derrida investigates the philosophical background of these considerations by showing how Rousseau thinks technics, and especially the primary human technics of language, as an originary supplement. Technics and language are supplements, for they hardly belong to the naked essence of humanity, but at the same time they are originary supplements, for there is no naked essence of humanity, the state of nature has never existed except as a necessary fiction, and the supplements of technics and language are there from the start. The human being has no proper essence: his/her essence is his/her perfectibility, which goes as far as the capacity of producing his/her own essence especially through education, which can be interpreted as a technique of body and *esprit* (spirit). To sum up, Derrida reads philosophical anthropologies in which the essence of the human being is technics—and this is why there is no essence of the human—but he does not make an anthropological hypothesis of his own: this is what differentiates him from Stiegler.

Derrida then applies these general principles to the specific technics of writing. His analysis is well known (see, for instance, Derrida 1981, pp. 61–172 and Derrida 1982, pp. 307–330). For him, writing is not a means and cannot be reduced to the author's intention. It would not function *as* writing if it could not function also in the absence of its author (and of the signified). As such, writing *is* nothing else but the signifier, which functions like a machine that repeats the code that has been entrusted to it. It does not only conserve the signifier as such, but, as it passes on to new readers in new contexts, the signifier can also be interpreted otherwise or not be understood at all, and this is why it can also produce many kinds of misunderstandings and unexpected new readings. Quite like a technical object, according to Leroi-Gourhan and Simondon, it can be transferred to many different contexts in which it gets different uses. This does not mean that anything goes: this only means that writing, like any technique, enables both the loss of the original sense/use and the invention of new senses/uses. Writing is not an obedient means of the intention to signify, but on the contrary it puts this intention to risk. In *La Pharmacie de Platon* (*Plato's Pharmacy*, in Derrida (1981)), Derrida interprets writing as a *pharmakon*, which is at the same time a remedy and a poison, for writing at the same time helps memory and ruins it (for you may forget what you have entrusted to the archives of writing). The *pharmakon* of writing is for him the very prototype of the prosthesis that at the same time supports life and reveals its weakness. Derrida shows that "writing" is indispensable to us humans—meaning, *some* kind of writing, actually any trace of a signifying activity, for the term has reached a very general signification by now. We are dependent on such traces of meaning because we cannot be conscious about everything we think, not to speak about what other people think, and most of the time we need to think without being conscious of it. This happens through the *pharmakon* of signs, texts and archives that come to us from our past and from other people. Writing haunts us, and we are what we are only thanks to this haunting of the *pharmakon* of writing.

Derrida has developed these basic gestures in many texts on many different subjects: I will not repertory them here. Instead, I will indicate how his younger friend and colleague Jean-Luc Nancy has applied the same logic to his own heart implant in *The Intruder* (2008, originally published in 2000 as *L'intrus*). Describing his personal experience of having a heart implant, Nancy underlines that the prosthesis consisting in another person's heart is far from being a simple means of better health, that the body would incorporate without further ado. It has the ambiguity of the prosthesis such as we have just described it. On the one hand, the implant supports the life of the body that would otherwise die. But on the other hand, instead of becoming one with the receiver's body, the implant remains alien to it: its alienness is seen, for instance, in physical reactions of rejection, in psychological estrangement, in technological and social dependency of the medical institution, and so forth. Instead of restoring health, the implant induces a technically produced chronic illness: the implant is an ambiguous intruder that protects the body from one kind of death by introducing in it another kind of death.

Nancy's example is meant to illustrate the general logic of the prosthesis. The prosthesis is and remains an external thing incrusted in the living being. Even the most perfect of prostheses is not one with the body but a stranger within the body proper: "My heart became my stranger: strange precisely because it was inside" (Nancy 2008, p. 163). As Derrida might put it, it "haunts" the living being as a presence that is at the same time strange and intimate: it has the uncanniness (*Unheimlichkeit*) that constitutes the intimacy of the living being by rendering it stranger to itself (the theme of haunting is developed especially in *Specters of Marx* (Derrida 1994)). In Derrida's analyses, all kinds of technics function in a prosthetic manner: psychological techniques, material technical objects, total technological systems—all of them end by supplementing human life as its prostheses. They constitute it by alienating it from itself; they haunt it as artificial supplements that impress their strangeness on life.

At the same time, there is no pure life that would not be marked by the "haunting" of "prostheses", and humanity marked by technicity has no proper figure but appears at most as an infinite figurability. But even this is presuming too much. Like Derrida, Nancy refuses all forms of anthropology that claim to answer the question "what the human being is?" ("Qu'est-ce que l'homme"): as Nancy and his friend Philippe Lacoue-Labarthe affirm in their introduction to *Les fins de l'homme* (the Cerisy colloquium on Jacques Derrida in 1980), all forms of this question ultimately reflect the anthropocentric humanism that is produced by "the age of technics" (Nancy and Lacoue-Labarthe 1981, pp. 12–13). "It's not just a question among others concerning the human being: instead, it is the human being itself that becomes a question, and this happens for the fundamental—that is to say not accidental—reason that the thinking of the human being becomes thinking of the finitude, or to put it more rigorously, that the ontotheology of the Subject finds itself obliged to turn into an analytic of finitude" (ibid., p. 13, my translation). In order to escape the ontotheology of the subject, one ought to learn to ask *who*, and not *what*, a human being is. As Nancy shows more precisely in "Un sujet?," the philosophical question does not pertain to any *anthropological* determination but to the question of the *subject of philosophy*, which Hegelian philosophy brought to completion. What comes after the subject in this sense is not another kind of humanity—or for that matter anything like trans- or posthumanity—but a thinking of existence in terms of finitude. Finite thinking asks *who is there*, *is there somebody* (Nancy 1992, pp. 89, 92), it aims the somebody that cannot be reduced to any *type or kind* of humanity but that is always "strange" (bizarre), always already an exceptional, singular, plural existence (the construction of the notion of somebody, *quelqu'un*, is the object of *Être singulier pluriel* (Nancy 2000)). It is only the withdrawal that conditions any discourse as such, as Nancy has shown in his first books (*Discours de la syncope* (1975), *Ego sum* (1979), *L'impératif catégorique* (1983); see Morin 2016). Correlatively, Nancy thinks, "Our world is the world of the 'technical', a world whose cosmos, nature, gods, entire system is, in its inner joints, exposed as 'technical': the world of the ecotechnical" (Nancy 2008, 89). The world's technicity does not hide a truer world—nature, being—but it *is* the world's very way of being (cf Lindberg 2018).

This being said, how does technics contribute to existence? This question has been developed especially by Bernard Stiegler in his trilogy *La technique et le temps* (1994, 1996, 2001) (Stiegler 1998, 2009, 2010) where he develops an entire "existential analytic, which should accordingly be *interpreted in terms of the question of prostheticity*" (Stiegler 1998, p. 183). While Derrida's main interest is in the quasi-transcendental conditions of sense, Stiegler takes his distances from Derrida's quasi-transcendental approach by formulating, not exactly an anthropology, but an existential analytic that underlines the role of technicity. His theory of technics is original, for he complements the Simondonian theory of technical objects by interpreting technics as time—as the prosthesis of time. Technics is time because it is essentially memory: not the *genetic* memory that is based on our biology, and not the *epigenetic* memory that is based on our personal experiences, but what Stiegler calls the *epiphylogenetic memory* in which an entire community conserves its experiences. Technical objects are the carriers of this epiphylogenetic memory: they incarnate the memory of an entire society in such a way that individuals do not have to know personally the things that are "known" by the technical objects used by their technical communities.

Thinking technics as epiphylogenetic memory has two important consequences. The first one concerns *knowledge*. Every technical object contains some knowledge of how material reality functions. For instance, to continue an earlier example, a hammer "knows" and "remembers" how wood and stone react to being hit. The technical community's knowledge of the material world has been inscribed in the design of the object. Because of this, the user of the object does not need to know it anymore—at least not so precisely—but s/he can trust that the object knows it already. The classical theory of technics as a means of the user's intention took it for granted that the user is a conscious subject of his/her technical activity. Stiegler shows that it is actually the contrary: the user can very well ignore why the tool works, because the tool knows it already. Insofar as a fragment of knowledge is included in the epiphylogenetic memory of the object, it belongs to the community even if its members had lost this knowledge altogether. Technical objects carry a trace of human knowledge of the world, and this is why humans think and act by using these technical objects. When doing so, they can very well think and act without knowing it: not only the user *possesses* knowledge that s/he does not personally know, but furthermore s/he *is possessed by* knowledge and ideas designed by somebody else. So to say, s/he is *formatted* by it. For instance, every user of gunpowder does not need to invent it, every driver of a car does not need to invent the wheel, and the rest. But with the gun and the car they incorporate the entire associated milieu of these technical objects, and moreover, the know-how and the values of a gun-and-car society. Of course, the ignorance concerning science, technology and raw materials that are needed for the production of the technical object is then compensated by another type of knowledge, for instance, that of knowing how to drive a car or how to shoot with a gun. But even then, technics favour a certain ignorance (for instance, if you think what you are doing while you drive, you probably find it much more difficult than if you just drove without thinking about it). In order to illustrate the ignorance presupposed by technology, Stiegler personifies human technicity not by Prometheus, the father of fire, but by his absent-minded brother Epimetheus, whose forgetfulness, weakness and clumsiness made human beings needy of technical supplements in the first place.

The second consequence of the interpretation of technics as epiphylogenetic memory concerns the *common* character of technics (see also Stiegler 2004a, 2004b). By definition, technics are common, shared by many. For instance, a hammer is not universal, like a mathematical proposition, nor is it unique, like an artwork. A technical object instantiates a *type*, for instance a certain type of hammer, that belongs to a certain technical community and might even define it, like prehistoric communities are defined according to their typical stonework or ceramics. Technical systems that are similar all over the globe are a recent phenomenon,[7] and even they ought to be understood as planetary systems rather than as universal truths. The common character of technical objects also means that the subject of technical activity is not really the individual who wields the object: s/he could be replaced by another member of the community, because the subject of use is actually the *operating function* that has been inscribed in the object so that it produces the operator in its side (Deleuze and Guattari 2000). In this sense, the real subject of the technical activity is the technical object itself, or through it, the community, in the form of the common knowledge that has been inscribed in the object. This is actually another way of saying that in order to correctly use a technical object, its user has to agree with its use, yield to its way of functioning, give him/herself over to it and, so to say, *be used by* it.

This phenomenon belongs to all technical communities, but it has become accentuated with industrialisation. Especially Heidegger, on the one hand, and Frankfurt School Marxists, on the other hand, have shown how modern industrial technology has grown into a global system, which is not really used by humans anymore but which uses humans as their operators and even as their raw material (they are "human resources"). Leroi-Gourhan has shown that technological communities have never coexisted with ethnical and political communities but they have always overflown them—and in this sense often enabled emancipation from the narrow limits of ethnical communities. In the modern industrial system, the distance between technological and political communities is still more marked. This means, on the one hand, that planetary technology gives opportunities to flee from oppressive political systems. But this also means that planetary technics has become uncontrollable to all political systems, notwithstanding some lame efforts of international regulation. Stiegler is very interested in this: he underlines that diverse technics contribute to our subjectivation but we do not really control this subjectivation, and in this sense we are not exactly the subjects of our technical systems but only their elements. For instance, none of us have designed the Internet such as it is, but we use it, and even our ways of doing philosophical research are transformed by it, for better and for worse. This is why technics is always a factor of both alienation and emancipation. This is obvious in the industrial civilisation but it actually characterises technics in general. As users of technical objects and systems, we are necessarily "inauthentic" and "alienated": we do what *anybody* could do with this object, and this is why we do not exactly realise our personal projects.

Stiegler underlines the "inauthentic" and "alienating" effect of technics. He does not mean, however, that somewhere in the heart of the human there would a non-alienated, authentic part that would be unpolluted by technics. On the contrary, the human condition is indelibly marked by inauthenticity and alienation—but these are not the best terms to describe human technicity. Instead, it is better to say that the human being is *originarily technical.* To start with, this means that the human is not only formed but constantly giving and getting form. As Rousseau shows, there is no human nature, but the nature of man is precisely his lack of nature and the corresponding capacity of artifice and technics. In this sense, there is no *humanity* that could support *humanism* or its posterior modifications (transhumanism and posthumanism as enlarged forms of humanity). Furthermore, Stiegler underlines that technicity characterises the human being as an *originary supplementarity*. This must be understood in the sense of Derrida's readings of Rousseau in *De la grammatologie:* he showed that what is supplementary is always secondary to a projected origin, that is, to something to which the supplement would be a supplement. But nothing precedes the supplement: there is no human nature, no originary thing, no full presence, nothing as such, except simply the *withdrawing of origin* that is the heart of originary supplementarity. Stiegler develops this arduous logic by showing that at the heart of promethean force there is always an Epimethean weakness: "the originary de-fault of origin" (Stiegler 1998, p. 199).

Thinking technicity as the *essence* of the human being—and not as his *supplement*—would amount to thinking that the human being might be a technical construct or product. This is the underlying unarticulated assumption of transhumanism, although transhumanists rarely argue in terms of essences. This is also the critical point of the kind of posthumanism that ends by postulating a new posthuman essence. When technicity is thought as the essence of the human being, s/he is interpreted as a machine that can realise a given program but cannot account for change and novelty. On the contrary, when technicity is understood as originary supplementarity, the core of technicity is lack. There is technics because existents are deficient—impotent, ignorant, weak, incomplete, shortly, finite beings—and technics is their ever-changing art of coping with their deficiencies. For the same reason, each technical object or system lacks something, as well: technics are not definitive solutions but just provisory *pharmaka* with their own defaults. There are no total and final technical solutions, but the human beings always encounter new problems and invent new solutions. Without the finitude marked by the lack, there would not be invention, novelty and change that can shake habitual mechanisms.

It is difficult to understand lack as the source of novelty if one understands the term only in the negative sense of privation. The positive aspect of lack comes to the fore when one remembers that the technicity of the human being does not characterise what s/he *is* (a stable essence) but what s/he can *do* (a dynamic possibility). Technics is not a question of substances and attributes but of capacity and potentiality. What Stiegler calls *épimetheia* is not exactly the absence of some *thing*, but the impotency of *doing* something. Aristotle's favourite example of *techné* was the doctor's capacity of healing, of which Aristotle remarked that the doctor, having this skill as a potentiality (*dynamis*), can also not heal, not realise his capacity. After Stiegler, we might therefore follow Giorgio Agamben, who suggests that technical *capacity* of doing and not doing is accompanied by an *incapacity* of this choice, both capacity and incapacity being of the same order of *potentiality*, that might one day evolve into a quite different *techné* (Agamben 2006). But this is another story.

Therefore

What distinguishes deconstruction from transhumanism and posthumanism? In this article, I have used the question concerning technics as the criterion of the distinction between these different currents of thought, showing in what sense they have different ideas of the role of technics in the self-constitution of the human being.

Transhumanists rely on an instrumental conception of technics. They see technics as a means of the realisation of the aims set by the human being, in order to modify him/her/self intentionally. Furthermore, human mind and body can be used quite unproblematically as raw material of such intentions. The moral value of the aims can be evaluated, while the instrument itself, as well as its raw material, is neither moral nor immoral but merely more or less efficient and available.

Most posthumanists see technics in a more or less Foucauldian way as a *dispositif* that disposes of us even before we can dispose of it. In our epoch, this general anthropological *structure* has become so intense and striking that it has engendered an entirely new anthropological *situation*, in which human beings rely so much on new technologies that they actually blend and hybridise with them. One could therefore characterise posthumanism as an affirmation of the new anthropo-technological situation that comes to the fore today.

The core of the deconstructive approach of technics as prosthetics is the antagonistic quasi-dialectics between supplementarity and lack. It is not opposed to the Foucauldian idea of technical dispositifs but it pays more attention to the element of lack that cannot be mastered by the technical dispositifs but that on the contrary makes them possible. Prosthetic technics is at the same time an indispensable element of subjectivation and its alien element whose effects are analogous to those of the unconsciousness. We are haunted by technics well before being conscious of them and their implications. Technics is fundamentally ambiguous, not because it would be good or evil as such, but because it has its own becomings. This is why Stiegler underlines the political dimension of technics. Politics of technology is not a matter of being in principle liberal or conservative in regard to technics as a whole, but a matter of recognising and evaluating the kind of existence and co-existence that each technology tacitly imposes on its users. Democratic politics today should learn to evaluate the associated milieus opened by each technology.

It is easy to see the difference between transhumanism and deconstruction. Transhumanists rely on an idea of a disincarnated mind that freely uses a material body, both of which are fully present. Deconstruction discards the mind-body distinction and studies existence, instead, and interprets existence in terms of lack, negativity and finitude. The standard transhumanist narrative suppresses the lack that poststructuralism takes to be irreducible. Deconstruction, on the contrary, claims that only finitude can explain the very possibility of novelty and change, that is the very condition of transhumanism, too. On the contrary, the distinction between deconstruction and posthumanism is more complicated. The kind of posthumanism that puts technics at its centre rejects classical humanism by affirming some version of cyborg existence, as depicted, for instance, in the works of artists like Orlan and Stelarc. On the other hand, another kind of posthumanism—sketched, for instance, by Stefan

Herbrechter—attributes the definition of "human nature" as the originary technicity to Stiegler and Derrida and calls for a different "posthumanism without technology", instead. The posthumanism without technology points at a protohumanity "behind" the supplementary technicity theorised by deconstruction, at an inhuman ancestrality defined by mediality (Herbrechter 2014). It would seem, however, that cyborg existence, existence as technicity and existence as inhuman mediality are all definitions of pre/post/in/humanity, while deconstruction on the contrary points at a finitude of thought that contests all definitions of (-)human.

Notes

1. Identification of technology as the motor of the future evolution of humanity is the central thesis of transhumanism, as stated in the *Transhumanist Declaration* (More and Vita-More 2013, pp. 54–55). Technology is central also for a more general experience of posthumanity that does not identify itself with transhumanism, like for instance in N. Katherine Hayles' groundlaying book *How We Became Posthuman?* (1999). How technology contributes to the emergence of the idea of posthumanity, see also Callus and Herbrechter (2004, pp. 226–257). Posthumanism is a label grouping quite heterogeneous positions. The possibility of a technological overcoming of classical humanism is imagined firstly in Donna Haraway's famous "Cyborg manifesto: Science, Technology, and Socialist Feminism", in Haraway (1991). Other forms of posthumanism study the possibility of an animal or ecological overcoming of humanism. This article focuses on the first one of these two versions.
2. In what follows, my subject is what ancient Greek called *techne̊* and what modern French calls *technique* and German calls *Technik*. However, the English language has three words for it: technics, technology and technique. I will speak about *technics* in a maximal sense that includes technique as know-how, technics as technical objects, and technology as entire industrial systems.
3. For instance, the transhumanist Nick Bostrom identifies himself as a posthumanist in "Why I Want to Be Posthuman When I Grow Up?" in More and Vita-More (2013, pp. 28–53).

4. Like Heidegger, Derrida, Nancy and Stiegler are "antihumanists", not because they would defend anything like inhumanity, on the contrary, but because they criticise all theories that claim to know *what* is the humanity capable of justifying *humanism* and, furthermore, *what else* is invited into *enlarged forms of humanity* (this is how some forms of trans- or posthumanism understand their own task). Humanity and humanism are not rejected because of their *ethos* but because of the philosophical status of these concepts that leaves them capable of betraying their own *ethos*.
5. Why would immortality be self-evidently good? Generally people dislike the idea of their own death—but they forget the fact that a deathless society would have to be a childless society, as well, unless humanity wants to die of a horrible overpopulation and a total exhaustion of resources, instead. What is sadder than a world without the novelty represented by children? In popular culture, the curse of immortality is treated, for instance, in Anne Rice's *Interview with the Vampire*, which Neil Jordan has turned into a film.

   Why would painlessness be self-evidently good? Aldous Huxley's *Brave New World* describes a society free of pain—for the greatest sources of pain, thought and feeling, have been drowned in powerful psychotropes and shallow entertainment.

   Why would intelligence be self-evidently good? Generally transhumanists take intelligence to mean tireless, flawless calculations. A computer does this better than man—but as for instance Hayles (1999, pp. 194–207) has shown, there are many other forms of intelligence that cannot be reduced to computation (sensibility, sociality, play, etc.).
6. Transhumanists generally take as granted that each person should be free to do whatever s/he likes with her/his own body and mind. This is defended by analytical bioethicists who insist that the state should not forbid the individuals' projects of self-transformation (Buchanan et al. 2000). This is precisely why today's transhumanists claim to defend the very opposite of twentieth-century totalitarian eugenics: their desire to enhance theirs and their future children's lives is precisely not about state controlling natality and education but about individuals choosing them differently. At first glance, this may seem as irrefutable as the necessity of freedom of conscience: one ought to be free to enhance one's body and mind as one sees fit (plastic surgery, gender reassignment, learning drugs, weird brain stimulators, etc.). But the case is much more complicated than this, because the freedom of enhancement is precisely not claimed

for oneself only but for one's children. Ordinary birth control and even embryo selection are not transhumanist. On the contrary, the genetic manipulation of the embryos is, especially since the invention of the genetic scissors CRISPRCat9, which make the fabrication of chimaeras a less *chimeric* idea. Genetic manipulation is not the proper autonomous consented choice of the future child, but of the adults who impose their will on the child's very body. Starting from the moment when a future child is the result of other people's projects, does it really make a difference if it is the state's project to have better workers or the parent's project to have the smartest children of the neighbourhood? I admit that quite cynically I doubt the intelligence of average parents as much as the goodwill of average states.

The most well-known critics of transhumanism have paid attention to the ethical and political problems connected to this problem in particular. Andrew Pilsch has labelled "bioconservatives" the critics who have warned against the consequences of transhumanism when it becomes a new eugenics (Pilsch 2017, p. 65). He names Francis Fukuyama, but one could have named for instance Hans Jonas (1985) or Jürgen Habermas (2001) as well. The term "conservative" is of course misleading, especially because it suggests that critics of modern consumer eugenics are critical because of religious reasons, while it is actually most often the other way round. On the contrary, critics like Jonas or Habermas warn against both consumer and state eugenics, not because they sanctify the present-day human genome, but because eugenics jeopardise the freedom and the autonomy of the child. Be it as it may, this debate on values is by nature without an objective solution: it is about values.

7. Nonetheless, this does not mean that technics would bear the stamp of the ethnical community of their invention. Stiegler quotes Leroi-Gourhan who shows on the contrary that similar techniques have been invented at the same time in many places: they reflect a certain relation between man and matter rather than the inheritance of the idea of one genial inventor (Stiegler 1998, p. 63).

References

Agamben, Giorgio (2006). *La Puissance de la pensée. Essais et conférences*. Paris: Rivages.

Bennett, Jane (2010). *Vibrant Matter. A Political Ecology of Things*. Durham and London: Duke University Press.

Besnier, Jean-Michel (2012). *Demain les posthumains. Le futur a-t-il encore besoin de nous?* Paris: Pluriel.

Braidotti, Rosi (2013). *The posthuman*, Cambridge and Malden: Polity Press.

Buchanan, Allen, Brook, Dan W., Daniels, Norman, and Wikler, Daniel, (2000). *From Chance to Choice. Genetics and Justice*. Cambridge: Cambridge University Press.

Callus, Ivan and Herbrechter, Stefan (2004). "Extroduction. The Irresistibility of the Posthuman: Questioning 'New Cultural Theory'". In *Discipline and Practice. The (Ir)resistibility of Theory*. Edited by S. Herbrechter and S. Callus. Lewisburg: Bucknell University Press, pp. 226–257.

Deleuze, Gilles and Guattari, Félix (1987). *A Thousand Plateaus. Capitalism and Schizophrenia. Translated by Brian Massumi. Minneapolis and London: Minneapolis University Press*. Deleuze & Guattari (1972/1973). *Capitalisme et schizophrénie, L'Anti-Œdipe*. Paris: Minuit.

Deleuze, Gilles and Guattari, Félix (2000). *Anti-Oedipus. Capitalism and schizophrenia*. Translated by Robert Hurley, Mark Seem and Helen R. Lane. Minneapolis: University of Minnesota Press. Deleuze & Guattari (1972). *L'Anti-Œdipe*. Paris: Minuit.

Derrida, Jacques (1976). *On Grammatology*. Translated by Gayatri Chakravorty Spivak. Baltimore and London: Johns Hopkins University Press. Derrida (1967). *De la grammatologie*. Paris: Éditions de Minuit, coll. Critique.

Derrida, Jacques (1981). *Dissemination*, translated by Barbara Johnson. London: The Athlone Press. Derrida (1972). *La dissémination*. Paris: Seuil.

Derrida, Jacques (1982). *Margins of Philosophy*. Translated by Alan Bass. Chicago: Chicago University Press. Derrida (1972). *Marges de la Philosophie*. Paris: Minuit.

Derrida, Jacques (1992). "Force of Law: The Mystical Foundation of Authority". In *Deconstruction and the Possibility of Justice*, edited by Cornell et al. London: Routledge, 1992. Derrida (1994). *Force de loi*. Paris: Galilée.

Derrida, Jacques (1994). *Specters of Marx, The State of Debt, the Work of Mourning, and the New International*. Translated by Peggy Kamuf. New York and London: Routledge. Derrida (1993). *Spectres de Marx*. Paris: Galilée.

Derrida, Jacques (2008). "Letter to a Japanese Friend". In *Psyche: Interventions of the Other Volume III*. Edited by J. Derrida and Translated by Peggy Kamuf and Elizabeth G Rottenberg. Stanford: Stanford University Press. Derrida (1998). *Psyché, Inventions de l'autre*. Paris: Galilée.

Foucault, Michel (2005). *The Order of Things*. London and New York: Routledge. Foucault (1966). *Les mots et les choses*. Paris: Gallimard.

Habermas, Jürgen (2001), *Die Zukunft der menschlichen Natur. Auf dem Weg zu einer liberalen Eugenik?* Suhrkamp: Frankfurt am Main.

Hagner, Michael and Hörl, Erich (editors) (2008). *Die Transformation des Humanen. Beiträge zur Kulturgeschichte der Kybernetik*. Frankfurt am Main: Suhrkamp.

Haraway, Donna (1991). "Cyborg Manifesto: Science, Technology, and Socialist Feminism in Twentieth Century". In Haraway (1991). *Simians, Cyborgs, and Women*, New York: Routledge, pp. 149–182.

Hayles, N. Katherine (1999). *How We Became Posthuman. Virtual Bodies in Cybernetics, Literature and Informatics*. Chicago and London: The University of Chicago Press.

Heidegger, Martin (1977). *The Question Concerning Technology and Other Essays*. Translated by William Levitt. New York: Harper & Row.

Herbrechter, Stefan (2013). *Posthumanism: A Critical Analysis*. London: Bloomsbury.

Herbrechter, Stefan (2014). "Unpublished Keynote Address Given at the 'Oh-Man, Oh-Machine—the Politics and Aesthetics of Posthumanism' Conference in Tel Aviv 15–20 May 2014)", available on https://www.academia.edu/13201635/Posthumanism_Without_Technology_or_How_the_Media_Made_Us_Post_Human, consulted on 24 September 2018.

Hottois, Gilbert (2014). *Le transhumanisme est-il un humanisme ?* Bruxelles: Académie Royale de Belgique.

Jonas, Hans (1985). *Technik, Medizin und Ethik*. Frankfurt am Main: Suhrkamp.

Kurzweil, Ray (2005). *The Singularity is Near. When Human Transcends Biology*. Penguin Books.

Leroi-Gourhan, André (1993), *Gesture and Speech*. Translated by Anna Bostock Berger. Cambridge, MA and London: The MIT Press. Leroi-Gourhan (1946 and 1965). *Le Geste et la parole 1. Technique et langage*, et *Le geste et la parole II. La mémoire et les rythmes*. Paris: Albin Michel.

Lilley, S. (2013). *Transhumanism and Society*, SpringerBriefs in Philosophy, DOI: 10.1007/978-94-007-4981-8–1.

Lindberg, Susanna (2016). "Derrida's Quasi-Technique". *Research in Phenomenology* 46, pp. 369–389.
Lindberg, Susanna (2018). "Onto-Technics—in Bryant, Harman and Nancy". *PhaenEx, Journal of Existential and Phenomenological Theory and Culture* 12, no 2, pp. 81–102.
More, Max and Vita-More, Natascha (eds.) (2013). *The Transhumanist Reader*. Oxford: Wiley-Blackwell.
Morin, Marie-Eve (2016). "Translator's Introduction to Jean-Luc Nancy". In *Ego Sum*. New York: Fordham University Press, xv–xxviii.
Nancy, Jean-Luc (1992). "Un sujet?". In *Homme et sujet. La subjectivité en question dans les sciences humaines*. Edited by A. Michels, J.-L. Nancy, M. Safouan, J.-P. Vernant and D. Weil. Paris: L'Harmattan, 47–114.
Nancy, Jean-Luc (2000). *Being singular Plural*. Translated by Rebord D. Richardson and Anne E. O'Byrne. Stanford: Stanford University Press. Nancy (1996). *Être singulier pluriel*. Paris: Galilée.
Nancy, Jean-Luc (2008). "The Intruder". In *Corpus*, Translated by Richard Rand. New York: Fordham University Press, 161–170. Nancy (2000). *L'intrus*. Paris: Galilée.
Nancy, Jean-Luc and Lacoue-Labarthe, Philippe (editors) (1981). *Les fins de l'homme. À partir du travail de Jacques Derrida*. Paris: Galilée.
Neyrat, Frédéric (2015). *Homo labyrinthus. Humanisme, antihumanisme, posthumanisme*. Paris: Éditions Dehors.
Pilsch, Andrew (2017). *Transhumanism. Evolutionary Futurism and the Human Technologies of Utopia*. University of Minnesota Press.
Ranisch, Robert and Sorgner, Stefan Lorenz (2014). *Post-and Transhumanism. An introduction*. Peter Lang.
Simondon, Gilbert (2012). *Du mode d'existence des objets techniques*. Paris: Aubier.
Stiegler, Bernard (1998). *Technics and Time 1. The Fault of Epimetheus*. Translated by Richard Beardsworth and George Collins. Stanford, CA: Stanford University Press. Stiegler (1994). *La technique et le temps I. La Faute d'Épiméthée*. Paris: Galilée.
Stiegler, Bernard (2004a). *De la misère symbolique 1. L'époque hyperindustrielle*. Paris: Galilée.
Stiegler, Bernard (2004b). *Mécréance et discrédit 1. La décadence des démocraties industrielles*. Paris: Galilée.
Stiegler, Bernard (2009). *Technics and Time II, Disorientation*. Translated by Stephen Barker. Stanford: Stanford University Press. Stiegler (1996). *La technique et le temps 2. La désorientation*. Paris: Galilée.

Stiegler, Bernard (2010). *Technics and Time III, Cinematic Time and the Question of Malaise*. Translated by Stephen Barker. Stanford: Stanford University Press. Stiegler (2001). *La technique et le temps 3. Le temps du cinéma et la question du mal-être*. Paris: Galilée.
"Transhumanist declaration (2012)" (2013). *The Transhumanist Reader*, see also https://humanityplus.org/philosophy/transhumanist-declaration/, last consulted in 21.9.2018.
Wolfe, Cary (2010). *What Is Posthumanism?*, Minneapolis and London, University of Minnesota Press.

Part III

Boundaries and Frontiers

Discourse Between Stefan Lorenz Sorgner and Paul Jorion on Nietzsche, Fascism and Moving Beyond Humanism: From Friedrich Nietzsche to Stefan Sorgner—The Short Path Leading from *Superhumanism* to *Metahumanism*

Paul Jorion

"Superman" or "Overman" as a recent fashion promotes to translate Nietzsche's *Übermensch*, to distinguish Him from a comic books' and Superhero films' character having gained prominence in the meantime.

The Superhuman is a superlative man who has dug through his own indomitable and Dionysian will-for-power a gap between ordinary men and himself comparable to that existing between men and other animals of a lower reign.

The Superhuman is not an enhanced version of us who will necessarily develop in a future to come: it is a project. His promoter dispelled any doubt in that respect:

P. Jorion (✉)
Department of Ethics, Université Catholique de Lille, Lille, France
e-mail: paul.jorion@univ-catholille.fr

P. Jorion (ed.), *Humanism and its Discontents*,
https://doi.org/10.1007/978-3-030-67004-7_4

> The problem that I set here is not what shall replace mankind in the order of living creatures (—man is an end—): but what type of man must be *bred*, must be *willed*, as being the most valuable, the most worthy of life, the most secure guarantee of the future. (*The Antichrist. Curse on Christianity* [1888] 1895: 3)

A project belonging to what family? The terms used "must be *bred*, must be *willed*" are devoid of any ambiguity: that family exists and goes under the name "eugenics".

Is there any displayable prototype for the Superhuman for us to see? The person of Friedrich Wilhelm Nietzsche himself who coined the label provides no doubt with sufficient matter to build the image and flesh it up.

Indeed apart from Nietzsche, and his enhanced portrait under the blown-up guise of Zarathustra, who else has made a viable candidate for being the Superhuman?

In Nietzsche's time, doctors were satisfied with the hypothesis that the madness that struck him at the age of 45 was due to his having reached the tertiary stage of syphilis, a venereal disease which, in those pre-penicillin days, was endemic with devastating effects.

Doubts have since been expressed about the aetiology, based on various factors, one being linked to the nature of the disease itself, which would probably not have allowed the philosopher to survive another ten years after his first undeniable psychotic episodes.

Peter Sjöstedt (2015) has recently proposed another explanation, which he convincingly supports with the help of numerous testimonies, based in particular on the philosopher's correspondence with his mother and sister—that of a Nietzsche who, over the years, drove himself mad through the immoderate use of alcohol and a panoply of hard drugs, a hypothesis supported by some of the philosopher's own remarks: "Let us take a step further: all those superior men, who felt themselves irresistibly urged on to throw off the yoke of some morality or other, had no other resource—if they were not really mad—than to feign madness, or actually to become insane" (*The Dawn of Day* [1881] 1911: 21).

Madness was in that case chosen as the deliberate strategy of the superior man: the *Übermensch*, the Superhuman.

However, Sjöstedt's hypothesis does not exclude the option of an evolutionary dementia, as Nietzsche mentions his morbid episodes on numerous occasions, hence the obsession with a return to health littering his essays: "...such a man requires one thing above all for his purpose, and that is, great healthiness—such healthiness as he not only possesses, but also constantly acquires and must acquire, because he is continually sacrificing it again, and is compelled to sacrifice it!" (*Ecce Homo* [1888] 1908), justifying, according to him, his evermore imprudent recourse, to the point of becoming unbridled, to alcohol—we're aware of his acquaintances being astonished at his capacity absorption—to opiates—the epithet *opiate addict* is appropriate in his case—and to chloral hydrate. Nietzsche states that "from out of a superabundance of life, the Superhuman combines those visions of the opium-eater, of madness and of the Dionysian dance" (in Köhler 2002: 226).

When Nietzsche sinks, he invents the Superhuman as a buoy to hold on to, as a version not only unscathed, intact, but, better yet, magnified of himself. Some 50 years later, an entire people sinking in its turn will cling to the same buoy of megalomaniacal delusion and spread satanic devastation over the world.

A major historical debate—far from over even today—is to what extent Nietzsche's views on the Superhuman have shaped the politics of Nazi exterminism. Due to the painstaking work of historians, and of Joachim Köhler in particular, we've got a much clearer representation of how the views of Nietzsche, Richard Wagner, Cosima Wagner's son-in-law Houston Stewart Chamberlain, so-called Hitler's John the Baptist, and Elisabeth Nietzsche's husband Bernhard Förster, coalesced to produce the theoretical framework of Hitler's exterminism.

Whether to exonerate Nietzsche himself from this snake pit is a question of personal inclination as he provided willingly weapons to that aim in the years he was under Wagner's spell. It was Cosima Wagner, née Liszt, who recommended Nietzsche not to say openly "Jews" about the vermin to be eradicated as he otherwise intended to say so more openly (Köhler [1997] 2000: 112), leading him to say "Christians", or even—following then a most tortuous and devious path of thinking—saying "Socrates" whenever Jews were meant.

It has now been fully documented, especially in Köhler's works, that it was Richard Wagner who had called in his days for a "grand solution" to the Jewish question unequivocally under the shape of a massacre (Köhler [1997] 2000: 110, 146–147, 176, 228). Nietzsche however—who was once submissively aping his manners—never managed to distance himself enough from Wagner's ranting to escape the orb of a planned genocide.

Adolf Hitler was enough of a devotee of Wagner—it is his attendance to a performance of Wagner's *Rienzi, der letzte der Tribunen* in his Austrian childhood town of Linz that decided of his call—that any speculation about the date of his decision to turn to the final solution as being either 1941 or 1942 is void: such was the way it had been planned by Wagner, first hinted at in his anonymous pamphlet of 1850, *Jewishness in Music*, later more explicitly stated in other publications (Köhler, ibid., 86–88). The sole question of the best time opportunity to be found had been left open.

In his book on Nietzsche's and Wagner's relationship, Köhler reports Goebbels' remark in his diaries that Hitler had once said: "That is why Nietzsche is inevitably far closer to the way we see the world than is Schopenhauer, for the task of philosophy is to simplify and to intensify life, not to cover it with a veil of pessimism" (Köhler [1996] 1998: 13). Köhler also reports, this time in his Nietzsche's biography: "In the First World War *Thus spake Zarathustra* was the favourite reading of the German troops in the trenches, and following the Kaiser, Adolf Hitler too found his image reflected in the mirror of the Superman, while Zarathustra's motto 'Glory be to the man of steel' was adopted by the SS" (Köhler [1989] 2002: 257).

In 1934, "Hitler had planned to build a commemorative Zarathustra monument to Nietzsche that would incorporate a statue of Dionysus donated by Mussolini, also an avid reader of Nietzsche's works" (Köhler [1996] 1998: 168).

In 1964, Erich Fromm (1900–1980), a psychoanalyst member of the Frankfurt School, had a special purpose when writing his groundbreaking *The Heart of Man*. The goal was to dispel once and for all the temptation for lazy and unimaginative commentators to try cutting short any debate about Hitler with a dismissive: "The man was mad! What else could be added?"

The tool to that aim was devising a new psychological category: that of the *malignant narcissist*. Here from *The Heart of Man*:

> The narcissistic person [...] reacts with intense anger when he is criticised. He tends to feel that the criticism is a hostile attack, since by the very nature of his narcissism he cannot imagine that it is justified. [...] If he *is* the world, there is no world outside which can frighten him; if he is everything, he is not alone; consequently, when his narcissism is wounded he feels threatened in his whole existence. [...] This fury is all the more intense because nothing can be done to diminish the threat by appropriate action; only the destruction of the critic—or oneself—can save one from the threat of one's narcissistic security. (Fromm 1964: 74–75)

This raises the overall question of in what degree of *tongue in cheek* we should ever be reading Nietzsche as he is the lonely, lonesome, reclusive figure, brushing up the portrait of a new type of humanity comparable only to what his bruised self was aspiring to become once again. Nietzsche along with others to his own image, the *daemonic* ones, in Stefan Zweig's vocabulary, fully absorbed with themselves only, a genuinely inverted picture of Man as depicted by Aristotle as the *zoon politikon*: the creature whose true constitution is to be social and thus inseparably bound to his fellow humans.

Here is what Zweig wrote about the brotherhood of Hölderlin, Kleist, Nietzsche, and like-minded souls, brothers in arms in the realm of excess and self-destruction:

> A power greater than theirs was working within them, so that they felt themselves rushing aimlessly through the void. In their rare moments of full awareness of self, they knew that their actions were not the outcome of their own volition, but that they were thralls, were possessed (in both senses of the word) by a higher power, the daemonic. [...] I term "daemonic" the unrest that is in us all, driving each of us out of himself into the elemental. It seems as if nature had implanted into every mind an inalienable part of the primordial chaos, and as if this part were interminably striving—with tense passion—to rejoin the superhuman, suprasensual medium whence it derives [...].
>
> They were nomads, vagrants, eccentrics, they were despised and rejected; they lived in the shadows [...]
>
> Hölderlin, Kleist and Nietzsche [...] were gamblers, staking their all with magnificent indifference upon the turn of a single card, to win or to

> lose a measureless prize; for the daemon loathes the tedious heaping up of petty gains in a saving box. (Zweig [1925] 2012: 11, 17 19)

Sorgner on the Concept of Human Dignity

If one considers Nietzsche's disconnected jottings on the "Superhuman" as the axiom to start from one's pondering, then Jaime del Val and Sorgner's 2010 *Metahuman Manifesto* should be regarded as a set of theorems derived from that axiom. To better understand what such theorems are about, the first sketch of the so-called Superhuman provided above was needed.

What makes the Superhuman any different from his brethren? His amorphous complexion, his vanishing boundaries. Here according to the *Metahumanist manifesto*'s authors:

> … the body as field of relational forces in motion and of reality as immanent embodied process of becoming which does not necessarily end up in defined forms or identities, but may unfold into endless *amorphogenesis*. Monsters are promising strategies for performing this development away from humanism.

Who is then the Superhuman that *metahumanism* has in mind? The answer has been given here: even if the monster does not provide the most likely central figure of the *Metahuman*, his mention means that the overall category should be wide enough that she or he is encompassed within it.

What now about Sorgner's claim that Nietzsche has undermined the concept of a human dignity and that the *Metahumanist* should follow him along that path? Let's examine carefully the reasoning's progress. Here is where it supposedly starts and ends, in Nietzsche's §115 of *The Gay Science*:

> The Four Errors.—Man has been reared by his errors: firstly, he saw himself always imperfect; secondly, he attributed to himself imaginary qualities; thirdly, he placed himself in a false order in rank in relation to the animals

> and nature; fourthly, he always devised new tables of values, and accepted them for a time as eternal and unconditional, so that at one time this, and at another time that human impulse or state stood first, and was ennobled because it was esteemed so highly. When one has deducted the effect of these four errors, one has also deducted humanity, humaneness, and "human dignity". (Nietzsche [1887] 1974: 174)

For a start, as Yunus Tuncel aptly observes in his review of Sorgner's *Menschenwürde nach Nietzsche: die Geschichte eines Begriffs*: "Nietzsche does not critique the concept of human dignity directly" (Tuncel 2011a: 1).

What does Nietzsche *state* in §115? Nothing more than that "human dignity" cannot be deduced from those particular four principles held together as they are false. Which does not mean that "human dignity" as such is therefore illegitimate and does not forbid that it can be derived travelling a different road.

There is ground indeed for claiming that "human dignity" is not the point of arrival of an ego-centred piece of logical reasoning but instead, for instance, a precept for communal living deriving from Man's nature of a *zoon politikon*.

> [Sorgner's] central critique concerning dignity is that it separates human beings categorically from other solely natural beings. (Tuncel 2011a: 9)

> From the perspective of my perspectivism, this also means that the prevalent conception of human dignity has no higher status in hitting truth as correspondence to reality than the conceptions of Adolf Hitler or Pol Pot. (Sorgner 2010: 242)

To this however someone may feel like replying: Good gracious old chap, in that case get rid of your perspectivism as it has failed the common decency test in the most dishonourable way! What further evidence would you still need?

Curiously Sorgner doesn't seem to envisage the self-evident alternative to getting rid of human dignity which is making it a particular variety of a wider dignity concept. That dignity should be allocated to many more than humans doesn't invalidate the principle in any way, let's just define indeed a concept of dignity whereof "human dignity" would be the one instance applying to Man.

> Sorgner: I enter the German discourses concerning human dignity to show that an essential element that is presupposed within the discourse is highly implausible, namely the presupposition that human beings are categorically different from other, solely natural beings such as animals. (in Tuncel 2011b: 5)

But on what grounds could "that human beings are categorically different from other, solely natural beings such as animals" possibly be considered a presupposition, since it can be experimentally established through the cross-fertility test which only succeeds with individuals belonging to the same species and to them only. Human beings are categorically different from other animals, and so are all animal species to each other.

Should we coin a notion of "gorilla dignity" to refer to the dignity we acknowledge to gorillas, it will necessarily be applied to gorillas only, unless we regard it impossible to claim "that gorillas are categorically different from other natural beings", which is absurd, unless one questions the entire possibility of setting up a biological taxonomy on an experimental base.

Of course when Sorgner reviews what prior philosophers have assigned as features justifying the concept of human dignity, expansion is difficult to anything but Man.

Thus when Cicero claims that human dignity lies in our essence, which is primarily rational, reason standing accordingly for memory, consciousness of the past, a capacity to plan for the future and to improve our living conditions, it would indeed be impossible to extend the category beyond the boundary of our own species (Sorgner 2010: chapter 2).

Or when Kant suggests a creature that is not rational cannot therefore be autonomous, a condition set for dignity (Sorgner 2010: chapter 5).

Kant's criterion does not allow any better than Cicero's criterion, more of a possible extension than does that derived from the *Homo Imago Dei* ones of olden: an immaterial soul, connected to reason, understanding, memory, free will, and immortality.

Those do not apply indeed to any other animal but us, and as Sorgner himself observes, it doesn't apply either to children or cranks, which would deny them then in the same way that so-called human dignity (Sorgner s.d.).

This being however an odd remark which one finds in different guises under Sorgner's pen: that a statement cannot be made if exceptions within its ordinary range can be found.

> For instance: There are humans who cannot feel physiological pain. Should they not count as persons? (Sorgner s.d.)

It all looks here as if Sorgner were not aware of the difference between the normal and the pathological, or claimed in some kind of Foucauldian way that the distinction between the two is arbitrary—therefore most probably his proposal to comprise the monster within the range—or possibly even better to regard the "monster" as the overarching category, of which to the so-called normal individual would be but a variety. Such a view would hold the advantage of making it possible maintaining the person of Friedrich Nietzsche right at the centre.

Sorgner's position has been well characterised by Tuncel when he writes:

> … how recent movements such as trans-humanism and post-humanism have responded to these issues; they have developed a different conception of the human with a critical distance to the concept of human dignity embedded in humanism. Some of the issues presented are abortion, active euthanasia, and genetic enhancement. The author shows how they can be supported by trans-humanist and post-humanist conceptions of human being; trans-humanism considers human beings as transitional and relies on technology as an agency of transformation towards higher goals and stages of being, whereas post-humanism foresees an entirely new human type that will overcome the type as conceived by humanism with its problems as exposed by the author under the rubric of human dignity. (Tuncel 2011a: 9)

At the same time this may very well look like having left the question hanging for having engaged it in a dead-end street.

So let us move one step backwards and start the investigation all over.

How else do we assign dignity?

For Schopenhauer it is because of sentience—or more specifically the ability to suffer.

For what reason do we assume that something, let's say a blob disembarking from a flying saucer, suffers? Because its behaviour is reminiscent of ours when we suffer. Do we demand from an animal that we believe is suffering that it confirms our intuition by saying explicitly so? No, for the same reason that we do not expect such verbal confirmation from a child who cannot speak yet.

The same with self-consciousness: the best confirmation of it is that the something, the blob, says: "I'm self-conscious".

We could therefore assign dignity to anything able to say "I", or that we would suppose on reasonable grounds that it would say "I" if only it could speak or send any other message we could hear. This is after all what we did at first contact with people whose tongue we failed to understand: we suspected that they were saying "I" in their own language.

But let us pursue. Why do we need the concept of dignity at all? Is it just a principle we need in order not to make suffer? But in that case wouldn't it be enough that we have a precept like "Do not make suffer"?

The answer to that resides in a number of other statements that Sorgner has made in that respect.

For instance, in "The Dignity of Apes, Humans, and AI": "Animals are not things but should legally be treated as things. Only human beings possess dignity". However, "a court in Argentina granted the status of a person to a great ape" (Sorgner 2019: 11).

"Being legally treated" as a human being as opposed to a "thing".

Or in "the idea that only human beings were given the immaterial divine spark is still legally valid. […] Every third German sees herself as a sceptic, naturalist, or atheist, which implies that the legally suggested understanding of animals is not shared by them" (ibid.).

As soon as "dignity" is being mentioned, the context seems to automatically become legal: the order prevailing in the courts.

Hence a thesis I will submit to the reader: "dignity" equates with not being *legally* treated as a thing. And to fulfil the necessary condition, only one thing is expected from us: that we behave responsibly, that we show responsibility, which is neither reason, understanding, memory, free will, immortality, autonomy nor the ability to suffer, it is something different: it is the ability to comply with the constraints defined as *imputability*, *accountability*, and *answerability*.

We Spontaneously Equate Dignity with Legal Responsibility

The case that Sorgner is referring to when he writes that "a court in Argentina granted the status of a person to a great ape" is that in December 2014, an appeals court in that South American country set a legal precedent by assigning an orangutan legal right to "life, liberty and freedom from harm" (Román 2015).

One recalls that Jean-Jacques Rousseau for one had already raised the issue of the dignity of the great apes, especially of the orangutan also called "pongo" in his days, in his *Discourse on the Origin and Basis of Inequality Among Men* of 1755:

> All these observations (...) make me doubt whether various animals similar to men, mistaken by travellers for Beasts without much examination, or because of some differences they noticed in their external conformation, or only because these Animals did not speak, would not be true Savage men, whose race, formerly scattered in the woods, had not had the opportunity to develop any of its virtual faculties, had not acquired any degree of perfection, and was still in the primitive State of Nature. (Rousseau III: 208)

An expert who, like Rousseau, reflects on what the "Natural Man" must be like looks into the issue of anthropoid apes and examines whether or not gorillas or orangutans are Savage men.

The facts are disturbing indeed as far as he is concerned:

> We do not see in these passages [wherein travellers report their contacts with great apes] the reasons on which the authors base themselves to refuse the animals in question the name of Savage men, but it is easy to conjecture that it is because of their stupidity, and also because they did not speak; weak reasons for those who know that although the organ of speech is natural to man, speech itself is not natural to him, and who know to what extent its perfectibility may have raised Civil Man above his original state. (Rousseau III: 210)

Although Rousseau does not settle the question, it is easy to guess where his feelings were heading:

> Hasty judgements, which are not the fruit of enlightened reason, are subject to excess. Our travellers indiscriminately see Beasts under the names of Pongos, Mandrills, Orangutan, the same beings whose divinities were made by the Ancients under the names of Satyrs, Fauns, Sylvans. Perhaps after more exact research we will find that they are men [that they are neither beasts nor gods, but men… (in the 1782 edition)]. (Rousseau III: 211)

Only true experts are in a position to decide on such matters:

> I say that when such observers (a Montesquieu, a Buffon, a Diderot, a Duclos, a d'Alembert, a Condillac, or men of that calibre [Rousseau III: 213]) affirm from such an animal that it is a man, and from another that it is a beast, they must be believed; but it would be a great naivety to refer to this, to coarse travellers, on whom one would sometimes be tempted to raise the same question that they try to solve on other animals. (Rousseau III: 214)

But where does this all leave us on the issue of the dignity of others, as in our own eye ours goes without saying?

In Robert Musil's *The Man Without Qualities* (or "without particularities" should we perhaps rather say), his hero, Ulrich, attends the trial of Christian Moosbrugger, the alleged murderer of a prostitute who had trailed him in the street. The verdict is death.

With Moosbrugger, Musil invites us to reflect on the more or less pronounced adherence of an individual to the acts she or he performs and of which the words she or he utters as a comment are the main components.

Here the passage where the points of view of the judge and the alleged murderer are confronted:

> This judge gathered all the elements, from the police reports and the accusation of vagrancy, into a single whole for which he charged the accused; but for Moosbrugger, it was only a series of entirely distinct incidents unrelated to each other and each resulting from a different cause, which was to be sought outside him, somewhere in the universe. In the eyes of the judge, his actions came from him, but in the eyes of Moosbrugger, they had rather happened to him like a flock of birds flying towards him.

"You always manage to shift the blame on to other people!" reproaches the judge to Moosbrugger, what one would usually translate into "you do not assume your responsibilities".

But by "responsibility" we actually refer to three different dimensions of what a court expects from a person summoned within its precincts: "imputability", "accountability", and "answerability".

What "imputability" summons is the mechanism which had been at work leading to the event justifying the court hearing. What is the factor without whose intervention the unfortunate event in question would certainly never have occurred? What we call the causal relationship.

We tend to say: "Everybody has always reasoned in terms of causes, and looking for responsibility means going after the cause of some event that happened". But in our culture, this is historically untrue: it is the notion of cause that derives from the notion of responsibility (Hart and Honoré 1956).

About natural phenomena, we used to say: "Knowing that it was obviously not a human being who made that terrible thing happen, who is the god who did it?"

And the fact is that we will still talk in court of an "Act of God" when something happens where we would normally expect there was a person in charge, but we're forced to admit that in this particular instance the person in charge was clearly not human. Thus we call it an *Act of God.*

And within the law, at that point, there is a reversal: it is the community as a whole which will take responsibility for the earthquake, the flood, and all things of such nature, the acts of a god who is not accountable.

Here still lie the roots of xenophobia and populism nowadays: the insistence of a number of us that there are no natural or "structural" forces, and we continue to necessarily attribute to particular ethnic groups, particular individuals, the responsibility that we had, so to say, the good taste, in ancient times to assign to gods who are outside our jurisdiction.

In the word "accountability", there is "account", as in "accountant". Accountability is our capacity to provide a rational account of what has happened, in terms that will make sense to a judge and a jury; the frame of reference here is the capacity for the person in the box to distinguish the false from the true.

"Accountability" refers to the ability to produce an articulate narrative. If we think of an AI, an intelligent machine, there will probably be no difficulty for it to be able to say how it attained a certain conclusion even if, like us, it would not have access to every one of the steps that it took within the neural network that carried out the reasoning.

In "answerability", there is "to answer", answering a question, or responding to it, where we find the etymology of "responsible". At stake here is the ability of the accused to understand why she or he has been summoned before a judge, why justice is considering putting her or him out of harm's way, and aims, as a side consideration, to ensure his future redemption. And to "answer" for her or his actions, by fully adhering to the status of the accused or the witness that is her or his at that moment, saying: "Yes, it is me indeed: I am the person in question!" In other words, the degree of her or his identification with the individual who is sitting in the box, and the elements she or he must then bring, with a view, the case arising, to lightening her or his sentence, to ensure that she or he will know how to make amends—should the verdict not be death. The frame of reference here is no longer truth or falsehood, but Good and Evil and the talent of each to be able to distinguish aptly between the two without being mistaken.

In order to impute, there must be evidence. As for the accused's ability to account, witnesses will corroborate or not what she or he says. As far as answerability is concerned: the disposition depends only on an accused person to understand for what precise motive justice is trying her or him.

Moosbrugger's ability to answer for his actions, he blows it up, even though the verdict has already been reached and the case seems to be closed. "Then the proceedings were concluded, and it was all over, wrote Musil, and he turned around, as the warders were already leading him out, he fought for words, stretched his hands above his head and shouted in a voice that shook itself free of his guards' grip: 'I am satisfied, even though I must confess to you that you have condemned a madman'".

Accountability in case of an accident caused by a robot? Yes, of course: the autonomous, driverless car that runs over a pedestrian is indeed the cause of the accident. The punishment here will be to put the robot's switch in the "off" position and to make its manufacturer or user liable, according to what the law will determine, supplemented then, for whatever ambiguities remain, by jurisprudence.

Ability to account for one's actions? Why not? An intelligent machine should be able to tell its own story, even if it has to embroider a little to improve the quality of the narrative, just as we ourselves do. But will it be sensitive to the precept that "it is wicked to lie"? I wouldn't bet on it.

As for the robot's ability to adhere to the person it claims to be, as for its degree of determination in being faithful to its word given when it has promised-sworn (and will then claim to be putting its honour at stake!), as for its talent to convince us of the sincerity of its remorse, guaranteeing its willingness to make amends in order to ensure its salvation, there, we will certainly ask to see!

In the meantime as far as animals, plants, rocks, landscapes are concerned, as they do not speak for themselves it is left to us to assign them dignity or not, alas, nothing can be done about it. My personal choice is to regard the universe as sacred, every atom in it. This spares me having to make fine distinctions between people I regard as likeable as myself and mere things, between friends to cherish and enemies to combat.

Conclusion

The person of Friedrich Nietzsche is the means that we've been given to constantly sharpen our ways of persuasion against the despairing, diabolical, objections that he opposes to us. Stefan Sorgner falls into the trap of taking him in earnest in whatever he wrote. For Nietzsche is not "the educator", the term he praised Schopenhauer with. Nietzsche has been playing a single part: that of a cunning devil's advocate.

So when I mention that the *Metahuman* elicits a feeling of *déjà vu*, I'm hinting at the fact that Nietzsche was not the first of our encounters with the *persona* of the narcissistic sociopath, or possibly even psychopath, although not under any malignant form in the case of our beloved philologist cum philosopher, but rather in the subdued variety, in the pitiful guise of the meek. Lo, meek only because of the multitudinous and unfortunate aggression of handicap, self-inflicted or not, on Superman's prototypical instance going under the name of Friedrich Wilhelm Nietzsche.

Indeed, if we follow the steps impeccably depicted by Peter Sjöstedt through Nietzsche's writings and correspondence, the emergence of the Superhuman in him follows in synch the development of his mental ailment. Thus a possible conclusion: *Metahumanism* is not to arise in the future as it has in truth accompanied humankind all along: the monsters that del Val and Sorgner have in mind are the cranks among us.

Sorgner's so-called *metahumanism* fails to exemplify a true alternative to humanism as it is nothing more than a trait present within humankind from time immemorial: what Zweig called the *daemonic*, the disruptive within the animal that we are, our untamed and disorderly attempts at solving existence's quandaries through self-destruction. True, there are sparks to be seen in the *daemonic* which are nowhere else to be seen within the human, just as volcanoes display an aspect of Nature absent from all its other manifestations. That doesn't mean however that the daemonic represents a genuine alternative to humanism, in the same way as volcanoes do not constitute a true alternative to Nature under its other and more benevolent guises.

References

Fromm, Erich, *The Heart of Man. Its Genius for Good and Evil*, New York: Harper & Row 1964

Hart, H.L.A. and Honoré, A.M., "Causation in the Law", *The Law Quarterly Review*, Vol. 72, 58–90, 260–281, 398–417, 1956

Köhler, Joachim, *Zarathustra's Secret. The Interior Life of Friedrich Nietzsche* [1989], New Haven, CT: Yale University Press, 2002

Köhler, Joachim, *Nietzsche and Wagner. A lesson in subjugation* [1996], New Haven, CT: Yale University Press, 1998

Köhler, Joachim, *Wagner's Hitler. The Prophet and his Disciple* [1997], Cambridge: Polity Press, 2000

Nietzsche, F., Ecce Homo [1888] 1908. http://www.nietzschesource.org/#eKGWB/EH

Nietzsche, Friedrich, *The Dawn of Day*, Translated by John McFarland Kennedy, New York: The Macmillan Company, 1911

Nietzsche, Friedrich, *The Antichrist. Curse on Christianity*, 1895 [1888], New York: Alfred A. Knopf, 1918

Nietzsche, Friedrich, *The Gay Science* [1887], New York: Vintage Books, 1974

Román, Valeria, "Argentina Grants an Orangutan Human-Like Rights", *Scientific American*, January 9, 2015, https://www.scientificamerican.com/article/argentina-grants-an-orangutan-human-like-rights/

Rousseau, Jean-Jacques, *Œuvres complètes III*, Bibliothèque de la Pléiade, Paris Gallimard, 1964

Sjöstedt, Peter, "Antichrist Psychonaut: Nietzsche and Psychedelics", Psychedelic Press Journal 12, 2015.

Sorgner, Stefan Lorenz, *Menschenwürde nach Nietzsche: die Geschichte eines Begriffs*, Darmstadt: WBG, 2010

Sorgner, Stefan Lorenz, "The Dignity of Apes, Humans, and AI" in *The Bioethics of the "Crazy Ape"*, Budapest: Trivent, 2019. https://trivent-publishing.eu/books/thebioethicsofthecrazyape/1.%20Stefan%20Lorenz%20Sorgner.pdf

Tuncel, Yunus, "Review of Stefan Lorenz Sorgner, *Menschenwürde nach Nietzsche: die Geschichte eines Begriffs* (Darmstadt: Wissenschaftliche Buchgesellschaft, 2010) 288 pp., ISBN 978-3-534-20931-6", *The Agonist*, Vol. IV, issue II, July–October 2011a

Tuncel, Yunus, "Stefan Lorenz Sorgner and Jaime del Val on the *Metahumanist Manifesto*. Interview by Yunus Tuncel", *The Agonist*, Vol. IV, issue II, July–October 2011b

Zweig, Stefan, *The Struggle with the Daemon* [1925], London: Pushkin Press, 2012

Dignity, Personhood, and the Sacred

Stefan Lorenz Sorgner

I do not wish to criticize Jorion in any detail for his reading of Nietzsche. Yet, a few points need to be noted briefly at the outset, as he employs them rhetorically to add inappropriate connotations to my version of metahumanism:

1. A more appropriate translation of the concept Übermensch than the word overman is the gender-neutral word overhuman. Übermensch does not have gendered connotations (Sorgner 2009).
2. To depict Nietzsche as an alcoholic is far from any solid scholarly understanding of Nietzsche, too. I merely wish to cite what Nietzsche himself stressed in the "Why I Am So Clever" chapter of Ecce Homo: "towards the middle of my life, I grew more and more opposed to alcoholic drinks".
3. Furthermore, it is misleading to hint at an association between Nietzsche's philosophy and the goals of the so-called Third Reich (Sorgner 2010b, section 9). Nietzsche was an anti-anti-Semite, he was

S. L. Sorgner (✉)
John Cabot University, Rome, Italy

P. Jorion (ed.), *Humanism and its Discontents*,
https://doi.org/10.1007/978-3-030-67004-7_5

> concerned with the power of creativity, and he was an anti-nationalist and a good European. In the same way, as Nietzsche's reflections are radically different from the ideals of the so-called Third Reich, the same applies to the reflections of transhumanists, even though, Habermas, too, identifies transhumanist reflections with Nietzschean breeding fantasies (Habermas 2001, 43–44). Jorion is in good company concerning his rhetorical misrepresentations. Both use these connotations rhetorically for presenting such beyond humanism approaches as monsters. This is widely shared humanist strategy, which has always had the tendency to represent the world as a fight between the good, white, rich, heterosexual men and the rest of the world. Jorion's intellectual fight against metahumanism is a continuation of this long-standing tradition.

Instead of playing eloquent rhetorical games, we should focus on the central philosophical issues. I need to correct Jorion's interpretation of my position concerning the criticism of human dignity, as I do not stress that humans are merely categorically different from other animals, but regularly specify further that the categorical difference applies to the ontological status (Sorgner 2010a). This is the decisive point of my criticism of the Christian and Kantian version concept of human dignity, for example.

> However, this anthropology has become less and less plausible from the end of the 19th century onwards due to insights put forward by Darwin, Nietzsche, and Freud. We no longer regard ourselves as being categorically separate from this world concerning our ontological status—we no longer hold that we have a material body and an immaterial soul. It is a more modest way of thinking, as it moves away from the traditional human hybris of possessing a categorically special status in the world. This does not mean that we do not possess special qualities. Learning a human language might merely be possible for human beings. However, animals also have such special qualities. Vampire bats manage to detect blood by means of infrared sensors. It is their special capacity. Animals, too, can have special abilities. Moving away from traditional anthropology implies that it is no longer plausible that only human beings participate in the immaterial realm, and are consequently attributed personhood, whereas all other

> beings are seen as things or objects. However, this is still the qualification which is legally valid in Germany and in many other countries. (Sorgner 2018, 6–7)

This movement away from a categorically dualist ontology has an enormous amount of implications. It was triggered by the realization that the realm of unchanging, eternal forms, God, and infinity, which can be accessed by means of rationality, has become implausible. Yet, it is this understanding which also implies that reason enables us to intellectually perceive the forms, which is what knowledge of the world is all about. We grasp an insight by intellectually accessing the realm of being and using words such that they correspond with the perfect forms of the immaterial, rational realm. By means of our senses, we cannot perceive any perfect equality. It can only be conceptualized rationally. We participate in both of the worlds, the immaterial and the material one, whereby the material one is permanently undergoing changes. The immaterial one is where Being is, and where the unchanging truths are, which are needed for there to be a correspondence between our true propositional judgements and the world. These truths have become implausible, and we realized that we are part of the one world which is permanently undergoing alterations in all respects. We can use words, form sentences, and make judgements. It is this capacity which used to be interpreted as us possessing and being gifted with the divine spark. Yet, the stories which people believed in to make this claim plausible have been undermined by the myriad of criticisms which took place during the enlightenment (Sloterdijk 1983). Different explanations have gained further plausibility.

Our capacity of having and being able to use a language is not the result of a divine spark, but rather of permanent struggles, and natural selections. We have developed a capacity of using a language, having thoughts, and making inferences, not due to a divine spark, but due to evolutionary processes, which implies that there is not one rational human nature, but rather that each human being possesses a slightly different concept of reason. Infinity, identity, and ideal are concepts which lack a correspondence to entities in the sensually accessible realm. However, using words usually works, that is, we usually manage to pragmatically convey insights such that speaking, and writing does what it is meant to do. When we ask for a loaf of bread, we usually get what we ask for. We have acquired language not during the act of fertilization when the fertilized egg gets connected with the immaterial rational soul, which is what the Catholic Church claims since the nineteenth century,[1] but when we get educated. Our culture and our parents are responsible for this development. They are responsible for and they bring about many of our upgrades. Our parents upgrade us with language. They turn us into cyborgs, that is, cybernetic organisms, which can be literally translated as organisms which have been steered, that is, upgraded or altered.

It is this revised anthropology which has myriad implications, whereby the consequences for our normative self-understanding and our attitude towards the notion of truth are particularly striking. Truth as correspondence to the world, as an epistemological goal, can no longer be meaningfully upheld, if all entities are permanently changing in all respects at all moments. How should a stable word be able to correspond to permanently changing entities? We are permanently exchanging poetic phrases which help us deal with the world. Still, each judgement remains an interpretation, whereby an interpretation does not imply that it has to be false but merely that it can be false. Perspectivism does not fall into the trap of the Cretan liar's paradox (Sorgner 2007). Nevertheless, it has striking implications, which I tried to highlight by means of the following poignant description, which Jorion cites to highlight the absurdity of my philosophical approach:

> Says Sorgner: "From the perspective of my perspectivism, this also means that the prevalent conception of human dignity has no higher status in hitting truth as correspondence to reality than the conceptions of Adolf Hitler or Pol Pot" (…).
>
> To this however someone may feel like replying: "Good gracious old chap, in that case get rid of your perspectivism as it has failed the common decency test in the most dishonourable way! What further evidence would you still need?"

On the contrary, I need to reply. Your approach is the totalitarian one, as you must affirm a concept of truth which claims universal validity.

However, Popper (2003) and Vattimo (1997, 37) have already convincingly demonstrated that such an understanding has totalitarian and violent implications, as such a strong notion of truth demands rigid, aggressive and dangerous implications. Such a strong concept of truth implies: Whoever is not with me is against me, is an enemy who needs to be defeated. We are the good ones, and you are evil. We possess the truth, you do not. You should be like us. It is this dualistic way of thinking, which is related to categorically dualistic ontologies, from which we have been moving away at least since Darwin and Nietzsche. Traditional Christian and Kantian humanistic approaches affirmed a division between the material world of change, emotions, and the body and the immaterial realm of stability, goodness, truth, and justice. The light is with the truth and so is everything and everyone who is white. Men are the rational ones who can access this realm of truths. God created each entity for a specific purpose, whereby the purpose of the genitals is to procreate. Any other use is unnatural and hence forbidden. Racism, sexism, heteronormativity, and speciesism are the cultural results of this aletheist way of thinking.

Aletheism brings about an enormous amount of violence against the monsters, and we are all monsters in one way or the other, as monsters are the ones who are very different from the norms, ideals, and perfections. People of colour, gays, lesbians, the handicapped, queers, transsexuals, and anyone else who is different from the norms of perfection which a categorically dualist way of ontological thinking have established. This is the reason why "*monsters* are promising strategies for performing this development away from humanism" (Del Val and Sorgner 2011, 1–4). Monsters provide us with alternatives, undermine the ideals, and reveal the power-structures related with specific discourses on cultural ideals.

There are people who consciously wish to become HIV positive. At the same time, I point out that most human beings identify an increased health span with a better quality of life. Anyone should have the right to live in his or her own monsterly ways, given that he or she does not directly harm another person. This is how proper plurality can be promoted. Incest among consenting adults, polyamory, or relationships between people of a vast age or power difference are morally justified if the people involved are consenting competent adults (Sorgner 2020b).

However, here, it is where the discourse gets tricky, as the concepts of consent and harm urgently need to be specified further. Is someone who consciously wishes to become HIV positive harming another person by doing so? If the person is part of a social structure which offers a public healthcare system, it can be argued that this is the case if that person after having become positive asks a physician for a drug to get treatment. In this case, the person's decision harms other people, as the prescribed drug will have to be paid for by all clients of the healthcare system. The money which will have to be spent on him cannot be spent on someone else, as all means of a public healthcare system are limited. This issue would not apply if no public healthcare system was available. The implications of this way of thinking are enormous, as it might have consequences for others who engage in risky sexual activities, extreme sports, or driving fast cars. Different kinds of risks apply to all of us. A rich society is able to afford the plurality of human interests. If a society is poorer, the situation is different. Many ethical issues come down to economic ones, which is the reason why the economic implications must always be considered, too. Furthermore, it needs to be reflected upon, if morally it makes a difference, if a state of affairs was realized as a consequence of a conscious choice (wanting to become HIV positive) or as an expectable side-effect of a risky activity (principle of double effect). These are the tricky issues which need to be addressed in this context and when we seriously consider the variety of implications which come about because of moving away from a categorically dualist ontology. However, Jorion's main issue is the concept of dignity, on which I will focus next, as he falsely claims that "Sorgner's so-called *metahumanism* fails to exemplify a true alternative to humanism".

A Non-speciesist Concept of Personhood

In many publications, I have presented alternatives to humanism (e.g. Sorgner 2020a). Jorion's approach, on the other hand, is devoid of pragmatically relevant content. The following passage clearly exemplifies this judgement:

> In the meantime as far as animals, plants, rocks, landscapes are concerned, as they do not speak for themselves it is left to us to assign them dignity or not, alas, nothing can be done about it. My personal choice is to regard the universe as sacred, every atom in it. This spares me having to make fine distinctions between people I regard as likeable as myself and mere things, between friends to cherish and enemies to combat.

All entities in the world are sacred. This seems to imply that all entities are persons or possess dignity. If this was the case, it becomes morally false to walk on grass, as you merely instrumentalize the grass, which undermines its status as a sacred entity. The same applies when you kill a mosquito, eat a Bistecca Fiorentina, or use a rat in a research lab. Such an approach does not have any pragmatically relevant content. Yet, to meaningfully deal with ethical issues, their applicability to the lifeworld needs to be taken into consideration. In contrast to Jorion's take, I stress the following.

It is implausible to identify personhood solely with human beings (Sorgner 2010b, 2020b). Human beings is a biological term, whereas personhood is a normative one, which implies a specific respect. The decisive question is what is the foundation for this respect. Traditionally, the dominant responses were that human beings possess rationality, autonomy, and free will and were created in the image of God. All of these foundations imply that all human beings deserve respect, as all of them possess the divine spark. The Catholic Church has specified in the nineteenth century that animation occurs with fertilization. Before the nineteenth century the current understanding of the Catholic Church as a gradual one, whereby male fertilized eggs receive the rational souls 50 days before this occurs to female offspring (40 days after fertilization for male offspring, 90 days for female offspring), according to Thomas Aquinas (Dunstan 1984, 41). The currently valid understanding of the Catholic Church implies that a human fertilized egg directly after fertilization deserves as much respect as a Nobel Prize winner or as you and me, even though they possess neither a nervous system nor a brain, which makes it impossible for them to suffer.

This is a highly implausible foundation for personhood. The normative concept of personhood should be separated from the biological one of human beings. An adult chimpanzee can suffer incredibly, too, which needs to be taken into consideration. In the argument "What Does It Mean to Harm a Person?" I explain in detail how to revise the concept of personhood, which ought to be founded on the capacity of suffering, as this is a morally relevant capacity, rather than in the membership to a specific species. Consequently, all entities which possess similar levels of suffering ought to be attributed with the same level of personhood, and together with this moral respect. It seems probable that this implies that many members of nine non-human animals which pass the mirror test should also be granted personhood. Passing the mirror test is an indication for possessing self-consciousness which again correlates with the capacity of suffering. A further indication for the capacity of suffering is the nervous system and the complexity of a brain structure. All these aspects are relevant for judging upon the capacity of suffering of an entity. Furthermore, it needs to be understood that there can be various types of suffering, that is, sensual, physiological, emotional, and cognitive suffering. When we consider cognitive suffering, we need to acknowledge that cognition might be sufficient for this type of suffering and that cognition can be independent of consciousness. Given that this is the case, it can be argued further that not only many members of non-human animals ought to be attributed with personhood, but the concept should also be in principle open to embodied AIs, that is, AIs who possess sensors, which enable them to cognitively interact with the world. This does not mean, by the way, that they need to possess consciousness. I regard it as highly unlikely that consciousness is a capacity which we can meaningfully attribute to silicon-based entities, at least in the near future. Yet, I do not exclude the possibility that it might occur sometime. Furthermore, I need to highlight that I am not claiming that currently available embodied AIs already possess the relevant capacities for being regarded as persons. This is not the case so far. What is important is that the concept of personhood should become a less speciesist one, as is already the case in Argentina, where an Orangutan was recently granted personhood, which had the consequence that the Orangutan had to be freed from the zoo.[2]

A further worry concerning my suggestion on how the concept of personhood ought to be revised comes from a different political interest group, which is hardly less dangerous than the aforementioned one.

Many critical posthumanists argue that sticking to the notion of personhood still implies being stuck in a dualist humanistic framework. It is important to stress that this is not the case. Here, it needs to be highlighted that the relational counter-suggestions presented imply the coming about of new paternalist structures, which is highly dangerous due to the totalitarian implications. I have explained this point in detail in a recently published article in German, which demonstrates the implications of a relational ethics with a case which occurred in China. There a boy was punished for a crime which was committed by his dad. The normative evaluation of the boy is dependent on the relational complex to which he belongs, that is, he is the son of his dad. This is a necessarily paternalistic implication of a relational ethics, which is an implication which needs to be avoided due to the illegitimate harm which it brings about (Sorgner 2020b).

In addition, the aforementioned argument by some critical posthumanists implies the same mistake which was committed by Jorion. Both counterarguments rely on a misunderstanding concerning the implications of moving away from a categorically dualist ontology. It does not imply that we need to entirely get rid of a dualist wording. It merely implies that a dualist wording has lost its ontological foundation. For pragmatic purposes, it seems important to rely on words, which are often coined in a dualist manner. The danger comes up when we associate a valid ontological and normative foundation with the words. The situation is an entirely different one if we realize and stress that the usage of words and norms is fictitious. Words and norms are created by human beings. They are contingent and subject to debate. This also applies to my reflections on personhood, which do not claim to be a new foundational reply to the question of normativity.

In addition, it needs to be mentioned that my alternative suggestion concerning personhood is not a dualist one either, as there are degrees of capacities of suffering. Consequently, we need to develop degrees of personhood, which clearly shows that this counter-suggestion is not a dualistic model. This, however, does not mean that I presented a fully developed new concept of personhood. This goal still needs to be realized (Sorgner 2020b).

The aforementioned reflections provide a path for further thinking, which particularly focusses on the achievement of negative freedom and the absence of constraint, as it is a wonderful achievement and one which we have been fighting for many centuries that we no longer get dominated by religious and political leaders concerning how to lead our lives. This is a precious achievement which we need to cherish. Personal flourishing gets promoted best if society allows us a great plurality of concepts of the good (Sorgner 2016). Of course, this does not imply that no limitations are needed. Raping another person is morally wrong and needs to be sanctioned. Limitations are important whenever one person directly harms another person. However, the concept of harm needs to be reflected on and clarified further. This is a much trickier challenge than is often taken for granted.

In addition, it is central that we do not get forced to be judged on an ontological basis to which we do not subscribe. Ontological judgements clearly need to be separated from legal issues. The German foundational law and many constitutions hold that only human beings possess dignity and animals ought to be judged on the basis of the object law. This premise has problematic ontological implications, which are inconsistent with the foundations of a liberal-democratic pluralistic society. Dignity is a concept which has descriptive and normative elements, and this is the reason why the above-mentioned judgement is morally problematic. Instead of employing concepts which have a descriptive dimension, we should try to focus on fictitious normative concepts only. If they are fictitious, they do not have any better foundational epistemological status than the worldviews by Hitler and Pol Pot. However, the epistemological basis should be irrelevant for legal issues. What is important is that negative freedom, the absence of constraint is a wonderful achievement.

We have been fighting for freedom for centuries.

We do not want any totalitarian regimes like the ones from Pol Pot and Hitler.

We need a system in which a radical plurality of personal flourishing can be realized.

We want a system, in which entities are being judged on the basis of morally relevant criteria and not on sexist, racist or speciesist prejudices.

I hope that my reflections on a concept of personhood which moves beyond a categorically dualist ontological foundation are a helpful path for realizing these goals.

Notes

1. "Pope Pius IX declared that animation begins with fertilization in 1869. He got rid of the distinction between a non-animated and an animated fetus, which used to be upheld by the Catholic Church until then" (see Sorgner 2020a, 218).
2. https://www.theguardian.com/world/2019/nov/07/sandra-orangutan-florida-argentina-buenos-aires, 12.11.2020.

References

Del Val, J., and Sorgner, S. L. (2011): A Metahumanist Manifesto. In: The Agonist 4(2), 1–4.

Dunstan, G. R. (1984): The Moral Status of the Human Embryo. A Tradition Recalled. In: Journal of Medical Ethical 1, 38–44.

Habermas, J. (2001): Die Zukunft der menschlichen Natur. Auf dem Weg zu einer liberalen Eugenik? Suhrkamp, Frankfurt am Main.

Popper, Karl R. (2003): Gesammelte Werke 5: Die offene Gesellschaft und ihre Feinde, Band 1: Der Zauber Platons. Mohr/Siebeck, Tübingen.

Sloterdijk, Peter (1983): Kritik der zynischen Vernunft. Suhrkamp Verlag, Frankfurt am Main.

Sorgner, S. L. (2007): Metaphysics Without Truth. On the Importance of Consistency Within Nietzsche's Philosophy. Rev. second ed., University of Marquette Press, Milwaukee.

Sorgner, S. L. (2009): Nietzsche, the Overhuman, and Transhumanism. In: Journal of Evolution and Technology 21(1), 29–42.

Sorgner, S. L. (2010a): Menschenwürde nach Nietzsche. Die Geschichte eines Begriffs. WBG, Darmstadt.

Sorgner, S. L. (2010b): Beyond Humanism. In: Journal of Evolution and Technology 21(2), 1–19.

Sorgner, S. L. (2016): Three Transhumanist Concepts of Human Perfection: In: Hurlbut, Ben, and Tirosh-Samuelson, Hava (eds.): Perfecting Human Futures: Technology, Secularization and Eschatology. Springer, Wiesbaden, 141–157.

Sorgner, S. L. (2018): Genes, CRISPR/Cas 9, and Posthumans. In: Sinaci, M., and Sorgner, S. L. (eds.): Ethics of Emerging Technologies: From Educating the Young to Engineering Posthumans. Trivent Publishing, Budapest, 5–17.

Sorgner, S. (2020a): What Does It Mean to Harm a Person?. In: HUMANA. MENTE Journal of Philosophical Studies, 13(37), 207–232.

Sorgner, S. (2020b): Konvergenz von Trans- und kritischem Posthumanismus. In: Wolf, P. (ed.): Transhumanismus, Posthumanismus und neue Technologien. Leipziger Universitätsverlag, Leipzig, 39–58.

Vattimo, G. (1997): Glauben—Philosophieren. Stuttgart, Stuttgart.

Truth? Still Breathing!

Paul Jorion

If it is regarded a rhetorical device (*artifice*) suggesting that a philosopher's aspirations at regaining not only perfect health but a superhuman quality might be related to his wretched, derelict, personal condition due to substance abuse (alcohol being but one along, more prominently, with opiates and chloral hydrate), should suggesting that the author of that hypothesis is likely to be heterosexual not be regarded as equally rhetorical? And should it not be rhetorical, what could it alternatively be?

But let us leave it to the reader to decide whether Sorgner's position and mine are in essence very close or seemingly widely discrepant only because we are deliberately and artificially making them remotely distant for the sole sake of polemical argument.

Let me concentrate, however, on one issue where it appears to me that the standpoints are in truth distant, mine being, as I see it, part of the Enlightenment's legacy, while Sorgner's being—no doubt to his amazement—still part of a historical obscurantism where humankind's efforts

P. Jorion (✉)
Department of Ethics, Université Catholique de Lille, Lille, France
e-mail: paul.jorion@univ-catholille.fr

P. Jorion (ed.), *Humanism and its Discontents*,
https://doi.org/10.1007/978-3-030-67004-7_6

at striving for knowledge were branded as blasphemous, meaning judged unredeemable for infringing on the deity's absolute and unsharable power at understanding.

What I refer to is when Sorgner claims, addressing himself to me:

> "Your approach is the totalitarian one, as you must affirm a concept of truth which claims universal validity. However, Popper and Vattimo have already convincingly demonstrated that such an understanding has totalitarian and violent implications, as such a strong notion of truth demands rigid, aggressive and dangerous implications. Such a strong concept of truth implies: Whoever is not with me is against me, is an enemy who needs to be defeated. We are the good ones, and you are evil. We possess the truth, you do not."

Allow me to focus on the claim that there is no merit whatever to "a concept of truth claiming universal validity". Such a statement, displaying a blaring and crass rejection of all achievements and victories of philosophy, deserves no less than an uncompromising rebuttal.

Let me proceed with order on this.

For Aristotle there were three types of statements that were true: incontrovertible reports of evidence from our senses, definitions, and conclusions of well-formed syllogisms.

The evidence of the senses produces truths insofar as we are not fooled by potential illusions. People we call *specialists*, *experts*, will agree that the evidence from our senses is reliable in this or that instance. There are propositions that we will be able to consider as true simply from such type of an experience we draw from the world: that it is, truly, in such and such a manner.

We will later develop a method—which will then be called the *experimental method*—to test whether in particular instances we are dealing or not with an illusion. We will produce hypotheses, work in the laboratory, invent instruments, to make sure that things are indeed the way they look to us as being.

This is what Plato referred to in his time and that was later called *adaequatio rei et intellectus*: correspondence between the mind's representation and how things really are.

See for this in Plato's *The Sophist*:

> The Stranger.—Now, the one of those two speeches that is true says, concerning you, what is, as it is.
> Theaetetus.—I cannot deny it.
> The Stranger.—And, obviously, the one which is false says what is, otherwise than it is?
> Theaetetus.—Yes.
> (Plato, *The Sophist*, 262 c–263 b)

Thus, a first type of truth that can act as a starting point: the adequacy of the word for that thing that the word depicts.

The second type of truth is truth by convention. If I say, "The fawn is the deer's kid", that statement is true simply due to an agreement having been reached through consensus within our culture that the expression "baby of the doe" can be replaced by the word "fawn". And from now on we can proceed on the basis of such a practical arrangement. In mathematics, a truth by convention is what is labelled an "axiom". These are starting points which have not been proven in any way or other, but which we will consider as valid starting points even if it is only by convention that they are true.

Finally, the third type of truth is provided by the conclusions of syllogisms. We'll be able to articulate with the evidence of our senses and with definitions, the conclusions of well-formed syllogisms, and we'll be able to advance in our understanding of the world by hopping from a syllogism's conclusion to another.

Having said that, there is a vast residue as far as persuasion is concerned, which opens up a crucial alternative dimension to world visions within human societies, to such an extent that we can ask ourselves nowadays if the truth that I've been depicting still exists. "Is there, one wonders, a new standard for everyday life other than that of truth?" And should there be indeed another standard than truth, is it of the order of truth or is it of an entirely different nature? Does the thing in question somehow deserve the status of "a kind of truth", to which additional components would need to be added in order to understand how it precisely works?

Let's imagine two people in a conversation where one says: "Personally, this story about the Earth being round has never managed to convince me. Truly I have always *believed* that the Earth was flat". The other then replies: "So, you too! This is something I have always *believed* as well and I'm particularly happy to realise that we both *know* the Earth to be flat!"

There is a possible transition, as we've just seen here, from "believing" to "knowing". Another example. One person says to another: "You know, So-and-so, I know him well and to be honest, he's an unadulterated rascal…". The other one then says: "Yes, I have no doubts about So-and-so being a scoundrel!" The conversation proceeds then from that agreement: "And since we both *know* that So-and-so is a scoundrel…". Something happened that caused these people who had used the word "believe" at first when they presented their own point of view, once they were convinced that each was thinking exactly the same as the other, to start talking no longer in terms of "believing" but in terms of "knowing".

What is at stake here is what semanticists refer to as "shared knowledge".

How does *shared knowledge* operate? When two people talk to each other, there is inscribed in the brain of each in the shape of memory traces, some information considered by them as settled data. When conversing, we're exploring what shared knowledge there is between the two of us: what is the overlapping part of the information that has been inscribed in our memory as a core of things we regard as knowing them.

As we move forward in the conversation, we are in a process of mapping this shared knowledge between us. At the same time, we remain in a negotiating mood. There is a part in what the two of us differ about that we are willing to jettison in order that we agree on the issue, and make it part of our new shared knowledge.

In terms of what we truly believe, as when we say, "I believe in God", the word "believe" might possibly mean that it is something we are less sure about, but in truth it is not: belief is something that we are precisely not ready to negotiate about. What we are willing to negotiate about is what we call our *knowledge*, because knowledge, as we perfectly know, can make progress: it is quite possible that we may have made a mistake and that we need to revise the information that we assumed was true.

And so, when we look at the genuine way that words like "knowing" and "believing" operate, we discover that we have quite often been off the mark. On "believing" Wittgenstein said the following in one of his reflections, written in 1938, the date explaining the "German airplane" in the example he uses:

> "Suppose someone were a believer and said: 'I believe in a Last Judgment' and I said: 'Well, I'm not so sure. Possibly.' You would say that there is an enormous gulf between us. If he said 'There is a German aeroplane overhead,' and I said 'Possibly. I'm not so sure,' you'd say we were fairly near." (Wittgenstein [1938] 1966: 53)

In everyday talk we keep negotiating. We negotiate on the basis of things we believe in, which are our core beliefs. We claim that we know for sure, still we negotiate in order to move ahead, and it goes without saying that in the relationship between the student and the teacher, a process is at work where the teacher tries to make grow the amount of knowledge that is shared between the two of them, by producing what for the student will be novel knowledge, so that she or he makes progress.

Negotiation is taking place in a permanent process but what is our method when negotiating? We do it from the perspective of more or less adherence to the words we utter. We're letting know to our interlocutor what it is that we are prepared to negotiate on and what there is no way that we would negotiate about. The person who says, "I believe in God" shows no intention of negotiating about the existence of God. While the person who says, "I heard someone tell me that God exists" is fully ready to negotiate on the topic. We use the words "believe" and "know", in phrases such as "I really believe", "I do know", "I know from having heard about it", or "I know for having experienced it", such are the words we use for making known our proclivity to negotiate on the views we claim are ours (cf. Jorion [1989] 2012: 222–225).

A recent change of algorithm by Facebook is very illustrative of what has just been said.

The social network had become aware that too many unproven conversation pieces were being disseminated, becoming points of focus in small communities where they became alive in the capacity of what those who reject them call sensationalist "conspiracy" tall tales.

Facebook modified its algorithm in reaction to what had emerged as interference of foreign agencies aiming to influence American and British voters, in the first case during the 2016 presidential election and in the second case with the "Brexit" referendum. Operators had intervened from an external outlet to encourage people who might sympathise with Trump's aims to vote for him and to discourage people who might be against him from voting at all. Similarly with Brexit: to encourage people to vote in favour of it.

Facebook realised that it had given too much leeway to ill-intentioned outsiders, through advertisements they may post, through the setting up of fake personae and groups, through calling, from abroad, for demonstrations at particular places at particular times.

The algorithm was accordingly modified: the potential impact of messages sent from abroad (Russia essentially) was reduced to encourage debate to remain within the groups that had been set up.

What happened then? A proliferation ensued of "local truths", in the likeness of those two chaps who would say: "We DO know the Earth to be flat!"

Facebook had now encouraged the emergence of purely local knowledge, the very opposite of the universally shared knowledge going under the name of "scientific knowledge".

Local truths being disseminated through *rhetorical* arguments exclusively, where whatever isolated instance is likely to be considered as conclusive: "I can assure you that I heard that someone…", "Indeed, you heard that someone…", "Yes, I heard convincingly that someone whose child had been vaccinated has reported that he had grown a second head…", and so on.

Operating within a small closed community only makes it easier to create a persuasive local truth. Of course, it is not the general truth emerging from the consensus of specialists, of recognised experts, but it functions as truth for the group of people involved who don't bother if it is based only on a purely locally produced type of knowledge.

Scientific discourse acts as a reference for all who care about finding out: it is open to every one of us to invoke it. Referring to science can be done by all of us in the exact same way as there are scientific protocols explaining how it works in an unambiguous way.

Now, in a society that has become tragically polarised, one sees on one side part of the population identifying with this type of universal knowledge, and one needs to add "who can afford to do so", and on the other hand, a share of the population coming up to oppose it in a clumsy manner with a rigmarole of alternatives remote from any scientific truth. Often its proponents are poorly educated, who would struggle to make a scientific type of reasoning, but come up anyway with something of a kind, whatever in fact their resentment has inspired them with, preferably in terms of wicked people rather than flawed structures. Their opponents may say that their views are prejudiced. But what is prejudice after all?

It could indeed be argued that it is only my own prejudice that leads me to regard the scientific method as unquestionable. The great Chinese philosopher, Zhuangzi (369–286 BC), asked: "What is it that allows us to move forward in a conversation? What makes it possible that we convince someone?" He replied: "Another can only be persuaded by me if I can hang something on to the prejudice he held to before we started speaking". Mine is, I admit, the scientific method.

Far from being "totalitarian", the search for truth, with a clear and unambiguous view of what we call so, constitutes the very backbone of what has become known and rightly revered under the name of *philosophy*.

Yes, some so-called philosophers do question that posit. However, clever and smart they might be and famous in some philosophical circles, I am among those claiming that they were no true friends of philosophy but members of a legion of obscurantists, that is, philosophy's actual foes. The very nature of their political affiliation serves in this respect as a cue.

References

Jorion, Paul, *Principes des systèmes intelligents*, [1989] Broissieux: Le Croquant, 2012.

Plato, The Sophist.

Wittgenstein, Ludwig, *Lectures and Conversations on Aesthetics, Psychology and Religious Belief*, Oxford: Basil Blackwell, 1966.

Enlightenment, Truths, and the Sciences

Stefan Lorenz Sorgner

Paul Jorion claims that his position is "part of the Enlightenment's legacy, while" mine is "still part of an historical obscurantism". By reading his as well as my own approach, it is up to the reader to decide which position is presented in a more clear, rigid, and comprehensible style, and which uses a more poetical, obscure, and hidden line of reasoning. Concerning the first part of the statement, I agree with him. His position is part of the Enlightenment. Yet, it needs to be noted that the Enlightenment is not a unified cultural period of time. There are several independent cultural movements which need to be acknowledged. Most noteworthy, a Kantian and a Nietzschean Enlightenment have to be recognized (Sorgner 2017). Jorion's approach is closely related to the Kantian version of the enlightenment, which affirms a rational human nature which is not accessible by means of the senses, and which enables us to grasp the truth in correspondence with the world. When Jorion refers to the "*adaequatio rei et intellectus*", it is this understanding of knowledge of the world which he refers to and affirms.

S. L. Sorgner (✉)
John Cabot University, Rome, Italy

P. Jorion (ed.), *Humanism and its Discontents*,
https://doi.org/10.1007/978-3-030-67004-7_7

The Nietzschean version of the Enlightenment, on the other hand, goes a step further and questions and reflects upon the origin of our rational human nature. This approach does not doubt that we have some kind of rationality, as it is clear that we can talk, form sentences, and make judgements. However, it takes evolutionary thinking seriously, and questions the origin of rationality which implies new limits of what rationality can achieve. We possess reason, but a unified concept of reason can no longer be made plausible. Reason is no longer the immaterial something to which all humans have access, and which we receive due to divine intervention at the moment when egg and sperm get united. This is at least the reading which the Catholic Church established in the nineteenth century. Instead, we all, in close cooperation with various other members of the human species, have developed our own very idiosyncratic concept of reason. The reason which I possess at the moment is different from the one you possess. However, it is still sufficiently similar that it does what it is meant to do. It enables us to communicate. Communication between us is not such that what goes on in my mind is identical to what goes on in your mind. However, the processes are sufficiently similar so that if I order and pay for a pint of beer in a pub, I usually get one. Words, judgements, and language are tools for making practical arrangements in the world, for establishing contracts, and for making plans for the future. All of these procedures do not guarantee that the connotations, which a word possesses in my mind, are the same as in yours, but by means of talking, reading, and writing, we achieve a sufficiently similar understanding of what words can mean so that language can fulfil its pragmatic function. The problem arises when we assume to associate truth claims, validities, and foundational and essentialist insights with our judgements (Sorgner 2020).

An interview which Cicero[1] conducted with the new realist Markus Gabriel and me in Spring 2019 comes to my mind, during which Markus Gabriel made the claim that human dignity, as it is affirmed in the German foundational law, really exists. The Big Bang has occurred 14 billion years ago, the earth came into existence about 4 billion years ago, the last common ancestors of humans and apes today lived about 6 million years ago, and homo sapiens came into existence about 400,000 years ago. However, the German foundational law including its concept of human dignity which came about less than 70 years ago possesses the only true valid ethical norm, and it took about 14 billion years so that the norm of human dignity was finally able to manifest itself in the German constitution.

Maybe, such a claim makes sense in the minds of those who have a specific religious upbringing, but if you take a critical and distanced stance, then it is far more plausible to judge that such a normative stance is merely the result of a specific interest group successfully defending their own prejudices such that they become culturally dominant for a certain period of time. Human dignity is a fiction which finds its support during a specific cultural historical period of time, as it has many implications and connotations which many Germans during the most recent past have found plausible. This is all. The central issue is the violence, which goes along with the claim of truth to which I have already referred to in the first reply, when I mentioned the totalitarian implications of all judgements which affirm an essentialist concept of truth in correspondence to the world.

This also applies to the claim that human dignity is such a truth. There is a myriad of highly problematic types of violence which go along with it; for example, it implies that any human being, merely by belonging to the human species, is infinitely superior to any non-human entity like an adult chimpanzee. Is this plausible? Should an adult chimpanzee, who can recognize him- or herself in the mirror, and has a developed brain and nervous system be legally treated as a thing, as it is the case according to the German foundational law? If it is the case that a human being merely by belonging to the human species possesses dignity, then a fertilized human egg without a brain and a nervous system is already such an entity, who possesses dignity. Consequently, this understanding of human dignity implies that a fertilized human egg deserves to be treated with the same respect as you and me, whereas an adult chimpanzee may be bought and sold, treated like an object and used for research which can make it suffer enormously. If someone claims that human dignity is a real truth in correspondence to the world, then these are the violent and counterintuitive implications. This is the reason, why Jorion's approach, too, has totalitarian and highly dangerous implications.

However, this is not all, as his philosophical stance is a rather fluffy pile of concepts, even though he accuses me of obscurantism. This comes out clearly, if the following reflections get dealt with. Firstly, Jorion affirms Plato's concept of knowing the truth, later on he claims that his method is the "scientific method". Firstly, he holds that by means of the intellect, he grasps the first principles, from which one can progress by means of deduction. Later on, he regards his own method for getting an understanding of the truth as an empirical one, which is also known as induction, whereby you start by means of various sensual observations, on the basis of which you can then form a general hypothesis. Is it the case that my reflections are obscure, or does this judgement not better apply to his thoughts?

I also talk about induction and deduction, but I do not claim that both methods are identical. In a world of permanent becoming in all respects at each moment, the possibility of an "*adaequatio rei et intellectus*" can no longer be conceptualized. A word is stable. A sensually accessible entity is permanently in the process of change, which renders any correspondence impossible. A Platonic form, on the other hand, can only be conceptualized in an immaterial realm with which we could not interact in the first place, and for whose existence we have no plausible reason at all. It needs to be realized that words do not correspond to eternal forms but are human made fictions which are used to better deal with challenges in the lifeworld. A word is a fiction which is used for communication, but no truth claim in correspondence with the world can be deducted from it.

This position does not imply any kind of obscurantism. It does not imply that one should not trust the sciences either, even though Jorion intended to stress that his approach follows the scientific method, whereas mine does not. This is not the case. In contrast to him, I clearly distinguish between induction and deduction. Knowledge cannot be gained by means of deduction, as this method presupposes premises which are the result of an "*adaequatio rei et intellectus*", which presuppose Platonic forms. Yet, language is merely a set of poetic fictions which enable us to deal with challenges in the lifeworld.

Knowledge cannot be established by means of induction either, but we merely come to realize certain correlations. These correlations are founded on the basis of empirical observations. If we have more observations, it is likely that the insight gained is more reliable, then if this is not the case. Yet, it must be noted that the working hypothesis which we develop on the basis of empirical observations might already have to be revised with the next possible observation. This does not mean that the working hypotheses which we gain in this manner are worthless. The contrary is the case. These working hypotheses do represent scientific truths. However, these truths are not insights which correspond with the world, but they are merely working hypotheses which have an empirical foundation. These scientific truths must not be underestimated. In many cases they are so reliable that they usually work. If this was not the case, we would not dare to get onto an aeroplane, as aeroplanes function on the ground of scientific observations.

Scientific truths are the best type of judgements we can get. A scientific truth is not knowledge in correspondence with the world, but they are pragmatic insights which usually work. This is as good as it gets. However, it is highly implausible to claim that we can gain any truths in correspondence with the world. Any such claim is not only implausible but furthermore highly dangerous, violent, and totalitarian, as such claims imply that no other judgement on the related issue is possible. Whoever is not with me is wrong and needs to be convinced, or maybe even fought against, as a judgement which claims to correspond to the world demands total validity. However, I have not yet come across any such statement, which can be regarded as valid.

I live in a world of poetic, contingent, and created fictions (Sorgner 2018, 2019). This does not mean that scientific insights are not to be trusted. Scientific insights are the best working hypotheses we can get, as they are founded on correlations between empirical observations. Even the intellectuals, who doubt the sciences usually trust them, as they often perform performative self-contradictions. Their actions contradict their intellectual judgements. They get on planes, even though they doubt the trustworthiness of scientific insights. Luckily, this does not apply to Paul Jorion. However, this is the case for many critical posthumanists, whose judgements can have highly problematic consequences.

In 2019, I invited the Nigerian transhumanist Leo Igwe as a keynote speaker to an 11th Beyond Humanism conference. There, he told the story of his father who had to go to hospital in Nigeria due to a serious illness, and he realized that his father did not get the treatment which he would have received in a hospital in Germany, where Igwe was awarded his doctorate. Igwe stressed that we Nigerians, too, wish to benefit from the scientific and medical insights which have already been established by researchers. Some critical posthumanists from the audience stressed the dangers connected with the medical companies gaining more and more influence in Nigeria, and with them destroying the local cultures of Nigeria. Igwe replied that the critical posthumanists, who made this claim, had gained their understanding of the local Nigerian cultures from Nigerian writers who had studied at the Sorbonne and Harvard, and who have then continued to live in Paris and New York as successful Nigerian writers, but whose novels are not being read in Nigeria at all. These novelists write for educated Western intellectuals, and not for actual citizens of Nigeria, who have very different worries. The local Nigerian citizens would love to have a good health system, a public health insurance, and the achievements which come along with scientific research in the field of medical sciences.

I agree with Igwe. There is not much which can be said about what is needed for living a good life. However, most persons regard an extension of their healthspans, as an increase in their likelihood of living a good life (Sorgner 2019). This insight is politically important, and ought to be considered much further than this is the case so far. Even though Jorion and I disagree with respect to the relevance and possibility of the truth in correspondence with the world, we both support the scientific method. This is already a helpful ground for further exchanges.

Note

1. https://www.cicero.de/kultur/transhumanismus-uebermensch-cyborgs-digitalisierung-technischer-fortschritt-silicon-valley-humanismus/plus, 12.11.2020.

References

Sorgner, S. L. (2017): Philosophy as "Intellectual War of Values". In: Blackford, R./Broderick, D. (eds): Philosophy's Future. The Problem of Philosophical Progress. Wiley/Blackwell, Hoboken/NY, 2017, 193-200.

Sorgner, S. L. (2018): Schöner neuer Mensch. Nicolai Verlag, Berlin.

Sorgner, S. L. (2019): Übermensch. Plädoyer für einen Nietzscheanischen Transhumanismus. Schwabe, Basel.

Sorgner, S. L. (2020): Wir sind stets Cyborgs gewesen. In: Herzberg, S/Watzka, H. (eds): Transhumanismus. Über die Grenzen technischer Selbstverbesserung. De Gruyter, Berlin/Boston, 2020, 61-82.

When Skin and Technology Intertwine

Hélène Jeannin

Introduction

The skin, one of the most visible elements of our body, is also one of the most misunderstood even though it is involved in all of the interactions with our environment. This vital organ serves as an envelope, barrier, and threshold, guaranteeing protection against external and internal aggressions (thermoregulatory function and the organism's immunity). Its relative fragility and the difficulty of reconstructing it despite recent medical advances (witness burn victims) make it both a mystery and a nerve centre for our survival.

As a subject of exposure or even decorum for the individual, human skin, under the influence of cybernetic thinking advocating the malleability of the body, may well become more than just a surface of contact with our environment. Because of its properties—particularly its plasticity and elasticity—it is currently the focus of a number of technical and

H. Jeannin (✉)
Orange/Department of Sociology and Economics of Networks and Services (XDLab/SENSE), Châtillon, France
e-mail: helene.jeannin@orange.com

P. Jorion (ed.), *Humanism and its Discontents*,
https://doi.org/10.1007/978-3-030-67004-7_8

scientific experiments. It is becoming much more than an interface with our environment: it is a living organ that is being intertwined with technology. This chapter offers an overview of current technological innovations that are introducing a host of new ways for skin and technology to interact, as well as the social issues such innovations raise.

When Transformation Begins with the Skin

The Skin, an Identity Under Construction

The skin has a look and shape that is unique to each individual. It can betray our intimate emotional or physical changes in a very visible way (blushing, pallor, sweat, etc.), and sometimes we cannot exercise any control over their manifestations, open to interpretation by others. Apart from the fact that skin can be an indication of an individual's ethnic and/or geographical origins, it is possible to decipher a person's age, physical condition, emotional state, and so on, by simply observing his or her skin, opinions that could be biased depending on our points of reference.

Skin, a potential source of discrimination, is an indicator of people's ability to live together. It can be a handicap or an asset, sometimes leading to dramatic consequences. There are more or less "favoured" skins depending on social and cultural instances, places, and times. Skin keeps track of an individual's personal history and we have a complex relationship with it: an identity can be built from a scar. The cult of youthful and luminous skin refers to the use of airbrush and concealment techniques so as to "cheat" on what real skin looks like. The velvety texture appears with striking clarity and excessive airbrushing maintains a sham. Member of Parliament Joe Swinson's 2011 complaint in England against Julia Roberts' and Christy Turlington's retouched photos has resulted in a law against the artificial and misleading representation of skin.[1]

The construction of our identity is also based on bodily norms, but these are not static throughout time. This is what historian Sylvie Steinberg shows from an analysis of gender transgression,[2] via the history of clothing over the centuries. Another illustration includes glasses which have gradually evolved from a held item (magnifying glass) to monocles, binnacles, or worn spectacles, then to incorporated devices (contact lenses), and finally to volatilised ones (refractive surgery). This secular evolution has transformed them into bodily objects and has made it possible to question the body-object coupling from a socio-historical perspective.[3]

Philippe Liotard reports in *Ceci est mon corps*[4] on the displacement of the body norm over the past forty years. Intrusive techniques for marking bodies initially confined to the underground and to homosexual, sadomasochist, and fetishist communities quickly spread throughout the Western world, becoming common techniques for elaborating on one's appearance. Individuals who use it may do so voluntarily and according to their own wishes. They experiment on the body with new techniques used in atypical projects (notably the implants invented by Steve Haworth in the early 1990s).

Tattooing is certainly the phenomenon that most clearly illustrates these practices, breaking off with the body markings of traditional societies. Now valued by sports stars such as David Beckham, and also many movie actors, it is becoming increasingly popular especially among the younger generations.[5] However, its main purpose in the past was to degrade the person, permitting recognition of his or her identification as a socially, mentally, biologically, or racially defective person. It was applied to slaves, criminals and prisoners, deserters, and soldiers, signifying their relegation to a lower rank and their ostracism of society. The literature echoed it: who does not remember Alexandre Dumas' Milady in *The Three Musketeers* and the wilt on her shoulder? Aristocracy and the political world in the eighteenth, nineteenth, and early twentieth centuries commonly practised it.[6] During the Second World War, American soldiers had pinups tattooed on their bodies thereby reminding them of the reasons for their fights.[7]

Whether a sign of virility or rebellion and widely associated with rock 'n roll or heavy metal music, tattooing defines a new visual culture[8] and carries meanings that have their roots deep in personal history. Emerging from quite marginal circles, the practice has spread to a very large scale. Influenced by tattoo shows the world over and a stronger and stronger sense of aestheticism, it no longer is an indication of society's rejection but rather a firm manifestation of the desire to distinguish oneself in a creative and very individual way.

This tattoo trend is being closely followed by research institutions and companies that may well draw much inspiration from it for their new technological inventions.

Skin: The Last Frontier of the Body?

The augmented body, cyborg, and other terms that seem to have come straight out of the cyberpunk literature of the 1980s have become a reality. This hybridisation[9] of a man with technology has called many theorists to question if or not, the skin is the ultimate body frontier. As far back as 1990, philosopher Donna Haraway, one of the pioneers of cyberfeminism through her memorable essay *A Cyborg Manifesto*, asked this question: "Why should our bodies end at the skin?"[10]

It is this idea that Bernard Andrieu took up again: "The excessive use of aesthetic surgeries has precipitated the body's cults into an indefinite quest for limits. This addiction defines a new type of clinical case: rather than borderline, individuals suffer from an internal absence of body boundaries."[11]

Rather than a natural, pre-determined entity, the body is increasingly considered as a malleable and porous construction.[12] Mike Featherstone (2000) sees body modification ranging from simple prosthetic devices to enhance body motor and sensory functions (e.g. spectacles), to the building of technological environments around the body, to the incorporation of technology into the body.[13] Some go further, forming an order that extends from the natural barrier of the skin to the military defences of a city or country. For example, in *Me++: The Cyborg Self and the Networked City*, William J. Mitchell[14] opens his argument by proposing a list of the successive layers that make up what he considers to be an already cyborg body, always connected with its environment, its tools, and the city in which it is located.[15]

Skin and Plasticity

Often, skin has been compared to fabric or clothing. This is how the expression "second skin" came about. The American television series *Nip Tuck*, and reality shows in the United States (*Extreme Makeover: Weight Loss Edition or The Swan*) accredit the idea that beauty is within reach of a scalpel, which simultaneously treats the ills of the body and mind. The bodily transformation is revealed to relatives and the general public who meet the final "result" with applause and approval, thereby glorifying the person and validating his or her choice.

However, recent research has rejected this idea of an inert membrane. A series of images in Dr Jean-Claude Guimberteau's[16] lecture "Strolling Under the Skin"[17] show a set of woven fibres and a micro-vascularisation, links whose threads stretch, move and recompose with each other, like a railway network where the nerve impulse is anything but linear. This interconnection, this dynamic mesh exceeds the most complex technology. The skin could well be another brain.

The Western cult of smooth, wrinkle-free skin was the subject of a caustic caricature in the film *Brazil* (1984) in which we see the face of a patient crushed like a simple cellophane film, stretched by the plastic surgeon in an excessive and unsightly way. Director Terry Gilliam will say about this sequence: "Every time I stay in Los Angeles in a hotel and take a magazine, half the pages are filled with advice on how to change your body and to do everything you dream of (…). One's modern dreams are corrupt, demanding immediate and short-term; they are about now, not about something really enriching, but on the contrary totally superficial."[18]

Fiction reminds us that transformation always begins with the skin.[19] It would be sufficient to just "change one's skin" in order to change one's identity or status, as in the Donkeyskin story (Charles Perrault). But the metamorphosis, in fantasy literature,[20] as in science fiction, remains a source of anxiety. It finds its climax in the transfer into the other's body. The body then becomes a revocable construction, a material that allows a new starting point (the film *Volte/Face*, 1997). The interchangeability of bodies would be a way of no longer dying but restarting one's existence.

We would change our skin as we change our clothes ("The Body", short story by Hanif Kureishi, the film *XChange*[21]). Attempts to appropriate the other's body may fail (*Dispossession* by C.H.B. Kitchin), or succeed (*The Story of the Late Mr. Elvesham* by H.G. Wells). The physical transformation, the "new skin" always has an impact on the mind and the life trajectory.

From these accounts, to which we could add organ transplants or prostheses, emerge the dominant hypothesis of the human body's infinite plasticity. *The Sopha* is an eighteenth-century Crébillon tale in which the author transports the intellectual and emotional data of his character into a reified interior, with the possibility for his soul to move from one piece of furniture to another. It anticipates Hans Moravec's idea of downloading the content of our brain onto a computer, in order to free man from his physical limitations.

Skin, a New Digital Interface?

Research Orientations

The ever finer integration of technologies into the human body has led to the resurgence of the term cyborg (a "cybernetic organism" used since the 1960s) to designate a man equipped with electronic devices. The extension of the definition of the cyborg to any living being and using tools, whatever they may be, has been proposed many times.

In the light of ongoing research, the idea of a man hybridised with flesh and technology is gradually taking shape. Technology is evolving to the point where devices can be integrated very closely to the bodies of individuals ultimately impacting their lives on a daily basis. From now on, electronic devices are no longer just in the pocket or in the hands. They can be inserted directly into (smart) clothing or skin. The *Internet of Things* is making a significant contribution to expanding the list of electronic devices connected to the body; however, these are part of a much broader system. Bionic prostheses that are grafted directly onto the body are sometimes controlled by a computer or directly by the brain.[22]

Let's quote the LifeHand2,[23] and many DARPA[24] projects: RE-NET,[25] HAPTIX,[26] NESD.[27] Implants are even more inextricably linked to the human body.[28] The process of video-entertainment technologies highlights a possible new conception of mobile phones and communication technologies in general—the transition from graphic interfaces and portable devices to immersive and embedded use.

The practice of enhancement takes place in an epoch of plasticity.[29] Research and development projects that use the skin as a support, substrate, interface, or source of information about an individual (by providing indications about his/her emotions, for instance) are the result of progress made in a number of fields, including biocompatible materials, sensor and processor miniaturisation, and nanotechnologies. They are giving rise to new kinds of tattoos (electronic ones) and sensors (connected body sensors).

Electronic tattoos are patches that can be stuck to the skin. They are flexible, contain an electronic circuit, and allow data to be transmitted wirelessly. They can be applied and removed easily, without breaking the skin. Unlike traditional tattoos, they are reversible, yet they bear all the hallmarks that have made tattoos such a success (beauty, transgression, and emotional association with personal experiences). They can serve as alternatives to the existing arsenal of data-exchanging body-based technologies (the *Internet of Things*, bionic prosthetics, implants, etc.).

Here are a few examples of the uses of electronic tattoos and connected body sensors:

- In healthcare and sports:
  - To indicate the state of a person's health[30] and certain biochemical parameters,[31]
  - To enable very detailed monitoring of patients (innovation by John Rogers and Yonggang Huang, University of Illinois, 2013),[32]
  - To analyse temperature rises and manifestations of emotions (Nanshu Lu's butterfly, University of Austin, Texas, 2013[33]).

- In authentication and security, the main selling point being greater convenience for users:
  - To unlock a smartphone without needing a password (Motorola Mobility patent, 2014, marketed by the startup VivaLnk with NFC tags[34]).
- In body decoration, beauty, and cosmetics:
  - To interconnect various systems on our skin: programmable colour-changing makeup, sensors that detect batting of the eyelashes or pursing of the lips, connected hair (see the work of Katia Vega at MIT[35] Media Lab).
- In other fields:
  - To detect facial expressions in order to control the TV, place a phone call, or control armchair height.[36]
  - To tattoo a sound wave (music clip, voice message) and then scan it to listen again through a dedicated application.[37]

Another example is the Duoskin project[38] carried out by Microsoft Research and MIT Media Lab in 2016. It illustrates the transition to a skin which, in conjunction with technology, becomes simultaneously:

- A user interface with the traditional features: buttons, cursors, and touchpads;
- An interactive digital screen, with data displayed and saved on the skin itself;
- A device or controller for managing objects remotely.

We can already glean on the web a prefiguration of what electronic tattooing could become if all existing components were combined in the skin (e-ink tattoo). The skin would become a screen revealing or hiding numerous patterns anywhere from a simple drawing to a chart of personal data (heart rate, glucose index, etc.) For instance, the display could appear on the forearm in real time.[39]

Potential applications are varied. The aim is to endow the skin with additional capabilities in terms of interactions, sensations, memory, and affect, thus turning it into a multipurpose *tool*. The possibilities are limited only by our imagination.

The selling point most often emphasised is a positive, humanist one, focusing on skincare, health, and body reconstruction. There is more than a hint of ambiguity, however. Often, one specific use is targeted and advertised (use for therapeutic purposes, for instance), but others are not necessarily excluded, hence the current surge in interest in topics surrounding the augmented body and human–machine interfaces.

These topics are particularly relevant at present, given the progress being made towards one day being able to implant user interfaces directly under the skin. For example, the possibility of permanently inserting electronic devices (tactile sensors, vibrators, speakers, microphones, wireless communication systems, etc.) beneath the skin to communicate with the outside world was already being studied back in 2012. The protocol, which involved testing such devices on human corpses, showed they worked once under the skin.[40] A study conducted in 2017[41] as part of the European *Horizon 2020* programme proved that the body markings specific to each individual (scars, tattoos, freckles, etc.) could be used as tactile control areas and be associated with remote devices and since they are easily identifiable, they are easily memorised.

Stakes and Players

The idea of a hybrid human being combining flesh and electronic technology is gradually becoming a reality through exploratory research and products that are already available or soon to hit the market.

These projects are often the result of massive investment strategies linked to public relations exercises by major players in the industry, the politico-economic world, the military industrial complex, and various other organisations. They develop alliances through research partnerships that guarantee transdisciplinarity, which is one of the leitmotivs of TCs (Converging Technologies). The question of embedded technologies, cyborg—in relation to that of posthumanism and transhumanism—also revitalises many programmes in the academic world.

Government agencies also support sometimes critical reflection projects, centred on hybrid technologies such as the Metabody Project, which is interesting because it is at the crossroads of the academic world and the citizen project. It is a project launched in 2013 and supported by the European Commission,[42] bringing together a think tank whose aim is to "develop a critical study of cultural homogenisation, social control and global surveillance in the information society and to develop new techno-cultural paradigms". Members come from several European universities, think tanks and laboratories, and the Advisory Board brings together leading academic and artistic figures interested in hybrid technologies such as Donna Haraway, N. Katherine Hayles, or Stelarc.[43]

Companies like Epicenter in Sweden[44] and New Fusion in Belgium have "equipped" their employees with RFID/NFC implants. Their promoter emphasises their practical usefulness (getting through security gates, making photocopies, ordering food without having to juggle multiple passes). However, they pose medical risks, may be perceived as intrusive, and are difficult to remove. Some also worry about possible abuses in biopolitical management[45] in terms of monitoring and surveillance of the population, and "technobiopower".

In a socio-technical system, the technological aspect must be considered only as part of an overall, integral system. The gradual elimination of cash from daily life, and the resulting constraints on individuals, must be taken into consideration. "Under the skin" technology raises the question of compatibility between flesh and metal, a problem that soon will be resolved with the advent of biocompatible materials.

In Scandinavia, Denmark and Finland have already eliminated almost totally, cash payments. These are replaced by mobile real-time money transfer applications[46] and the use of sub-dermal NFC/RFID chips is becoming wider and wider, to the point even, of paying one's train ticket with them.

Besides industrial players, there is an alternative culture. This consists of *Do It Yourself (DIY) Biology* or "body hacking". There are physical places offered by the community and also virtual places "online" where this can be performed and the tool kits are being sold. Most biohacking organisations have hatched in the United States under the 501(c)(3) status in which the U.S. federal tax code exempts them from the payment of federal income taxes;[47] this contributes to their development. Activities that are developing there have ramifications in ambitious projects that feed local initiatives and the economic impact is sometimes national. The biohackers network is increasingly structured around bio and body hacking conventions (BDYHAX). The inaugural convention in Austin, Texas, from January 27 to 29, 2017, was accompanied by an "announcement effect" for a prototype transdermal implant capable of welding to the skin.[48] These DIY spaces are based on the plasticity of living things. They are helping to spread new representations and practices in which skin is a raw material to be chiselled and sculpted.

This alternative culture operates much more underground and sometimes on the edge of legality. This is certainly a marginal phenomenon, but it sheds light on the relationships that coexist between counter diffusion and official mainstream diffusion. Some of these actors are thus linked to academic or economic fields; others enter into collaboration with companies for a broader commercialisation of the work they have sometimes carried out alone for decades, and so on. This panorama would not be complete without mentioning researchers such as Kevin Warwick (a reference for many hackers, despite operating somewhat on the fringes of the field due to his status as an academic[49]) and bio-art artists such as Orlan[50] and Sterlac.[51] They share their adventures with us in a narrative way, in the form of a personal quest and in front of the cameras.[52]

Neither should we forget the press and the museums, nor the cultural-based industries, or any kind of events (especially sporting) that are centred on the hybridisation of the skin with technology.

All these stakeholders in the vast field that is body modification are helping to change the social norms, a change that is vital to encourage the adoption of new practices. These new skin-based technologies, supported by pioneers with a sufficient media profile to influence behaviour, are accentuating body marking phenomena such as tattoos. Previously stigmatised, these are now valued and in certain circles are even becoming vital for social integration, or as the case may be, the total opposite of such. The body is increasingly seen as a malleable object to be continually reworked[53] and, as such, is becoming a popular target environment for sub-skin technologies.

Conclusion: Towards New Bodyguards?

Whether it is used as a source of information, a data storage medium, or an interface, skin is increasingly being used as a resource in a myriad of technological research projects.

In addition to these innovations, the skin's role as a barrier is also being disrupted. No longer does it prevent us from seeing inside someone's body! The exhibition "Body Worlds Decoded"[54] at the Tech Museum of Innovation in San Jose, California, invites the public to delve under the human skin and take an immersive, augmented-reality tour of the organs.[55] Lasers and imaging techniques also allow us to push our scopic drive and voyeuristic tendencies to their extremes.

The ability to break through the skin was previously restricted to the world of medicine and especially that of surgeons. Now, it is shared with a number of authorities and organisations: airports, police services, and private or state agencies that store biometric data. New usage practices are spreading in what is a more or less lawless area, the only assumption being that consent, be it expressed or just implied, is required. These practices dodge the question of risks and damage to health, as well as the principle of the inviolability of the human body.

They also encourage theoreticians and practitioners to challenge the idea of the skin as the body's ultimate barrier. The body has always appeared to have clear boundaries; the skin, its outer casing, marked the body's ultimate frontier. We are heading for a future in which the body is extended by miniaturised technology that interfaces it with the outside world constantly.

This opens up interesting possibilities in many fields (such as medicine), as well as undermining the notion of the hitherto clearly defined boundaries of a human being.

Notes

1. https://www.eonline.com/news/254745/julia-roberts-and-christy-turlington-ads-deemed-overly-photoshopped-banned-in-britain; https://www.asa.org.uk/news/why-real-beats-fake.html https://www.asa.org.uk/rulings/loreal-uk-ltd-a11-149640.html.

2. Steinberg (2001).
3. Veyrat et al. (2007/1 Vol. 1, n° 1).
4. Liotard (2015).
5. IFOP (Juillet 2010).
6. Borel (2006, p. 160).
7. Favre (mai 2012).
8. Fenske (2007).
9. Andrieu B. (2008).
10. Haraway (1990, p. 220).
11. Andrieu B. (2004/2).
12. "The Body and Technology: Discourses Shaping Consumer Experience and Marketing Communications of Technological Products and Services", Margo Buchanan-Oliver, Angela Cruz, *Advances in Consumer Research*, Vol. 36, 2009.
13. Featherstone, Mike (2000), "Body Modification: An Introduction", in *Body Modification*, ed. Mike Featherstone, London: Sage, 1–14.
    "Introduction," in *Body Modification*, ed. Mike Featherstone, London: Sage, 1–14.
14. Mitchell (2003).
15. "Consider, if you will, Me++. I consist of a biological core surrounded by extended, constructed systems of boundaries and networks […]. My natural skin is just layer zero of a nested boundary structure […]. My clothing is a layer of soft architecture, shrink-wrapped around the contours of my body. Beds, rugs and curtains are a looser assemblage of surrounding fabric, somewhere in between underwear and walls. My room is a sloughed-off carapace, cast into a more rigorous geometry, fixed in place, and enlarged in scale so that it encloses me at a comfortable distance. The building that contains it has a weatherproof exterior shell […]" (William J. Mitchell, *Me++: The Cyborg Self and the Networked City*, Cambridge MA: MIT Press, 2003, pp. 7–8).
16. http://www.guimberteau-jc-md.com/en/#.
17. http://www.guimberteau-jc-md.com/en/videos.php.
18. Audio commentary by Terry Gilliam on the DVD "Brazil."
19. *The Metamorphosis* by Frank Kafka (1915), *The Shrinking Man* by Richard Matheson (1954), *Thinner* by Stephen King (1984) or movies like *The arrival*, 1996, *The Fly*….
20. Marigny (2008).
21. http://www.imdb.com/title/tt0242150/.

22. Derian (2013).
23. https://robohub.org/tag/lifehand2/.
24. Defense Advanced Research Projects Agency, https://www.darpa.mil/.
25. Reliable Neural-Interface Technology.
26. http://www.darpa.mil/program/hand-proprioception-and-touch-interfaces.
27. http://www.darpa.mil/program/neural-engineering-system-design.
28. Jeannin (2016).
29. Vita-More (2010).
30. https://creapills.com/tatouages-sante-couleurs-biosenseurs-20170602.
31. https://www.sciencesetavenir.fr/sante/e-sante/un-biocapteur-pour-evaluer-la-sueur-en-temps-reel_101581.
32. https://www.youtube.com/watch?v=ABrHH9QoOHE.
33. https://news.harvard.edu/gazette/story/newsplus/nanshu-lu-engineer-of-electronic-tattoos-named-to-tr35/.
34. https://atelier.bnpparibas/en/life-work/article/digital-tattoo-unlocking-smartphone-easier.
35. MIT: Massachusetts Institute of Technology: http://web.mit.edu/.
36. Le numérique à fleur de peau : de nouvelles interfaces de communication.
37. http://www.huffingtonpost.fr/2017/05/10/tatouage-sonore_a_22079118/.
38. http://duoskin.media.mit.edu/.
39. https://www.youtube.com/watch?v=QMBX6nFg0RE ; what is e-Ink: https://www.youtube.com/watch?v=6NRT8E5-reM.
40. Holz et al. (May 5–10 2012) http://www.christianholz.net/2012-chi12-holz-grossman-fitzmaurice-agur-implanted_user_interfaces.pdf.
41. Bergstrom-Lehtovirta et al. (2017); http://www.sebastianboring.com/content/publications/prints/lehtovirta.CHI-2017.skinmemory.pdf.
42. http://metabody.eu/partners/.
43. http://metabody.eu/.
44. Cellan-Jones (29 January 2015).
45. In reference to Michel Foucault, who came up with the notion of bio-politics in the 1970s. It denotes power strategies which, in modern Western societies, are used on the body and the population with the aim of disciplining them, making them work, exploiting and controlling them.
46. For instance, Swish, Mobile Pay, and Swipp.
47. https://en.wikipedia.org/wiki/501%28c%29%283%29_organization.

48. BDYHAX: Austin's Inaugural Bodyhacking Convention to Feature New Projects From The Thought Emporium: http://www.prweb.com/releases/BDYHAX/01/prweb13162097.htm.
49. Formerly Professor of Cybernetics at the University of Reading, he is now Deputy Vice-Chancellor (Research) at Coventry University (UK): http://www.kevinwarwick.com/.
50. http://www.orlan.eu/.
51. http://stelarc.org/projects.php.
52. Declan (2015).
53. Useful points of reference include the many works of David Le Breton, particularly those on body markings (*Signes d'identité, tatouages, piercing et autres marques corporelles*, Métailié, 2002) and the skin (*La peau et la trace: sur les blessures de soi*, Métailié, 2013); as well as the research of Philippe Liotard on body modification: http://philippe-liotard.blogspot.fr/2011/12/modifications-corporelles-petite.html.
54. https://www.thetech.org/bodyworldsdecoded.
55. https://ph.news.yahoo.com/body-worlds-exhibition-embraces-augmented-162910933.html.

References

Andrieu, B. (2008). *Devenir hybride.* Nancy: Presses Universitaires de Nancy, coll. Epistémologie du corps.

Andrieu, B. (2004/2). Somaphore et corps biosubjectif. *Multitudes n°16*, pp. 59–69.

Bergstrom-Lehtovirta, J., Boring, S., & Hornbaek, K. (2017). Placing and Recalling Virtual Items on the Skin. *CHI '17 Proceedings of the 2017 CHI Conference on Human Factors in Computing Systems, May 06–11, 2017, Denver, Colorado, USA* (pp. 1497–1507). New York: ACM.

Brazil [DVD], by Terry GILLIAM, 1985, Embassy International Picture.

Borel, F. (2006). *Le vêtement incarné. Les métamorphoses du corps.* Paris: Calmann-Lévy, coll. Agora.

Cellan-Jones, R. (29 January 2015). *Office puts chips under staff's skin.* Récupéré sur BBC News/Technology: http://www.bbc.com/news/technology-31042477.

Declan, F. (2015). *The New Celebrity Scientists: Out of the Lab and into the Limelight.* New York: Rowman & Littlefield.

Derian, M. (2013). *Le métal et la chair. Anthropologie des prothèses informatisées.* Paris: Université Paris 1 Panthéon-Sorbonne (CETCOPRA). Ecole doctorale de philosophie.

Favre, C. (mai 2012). La pin-up US, un exemple d'érotisme patriotique. *Clio. Femmes, Genre, Histoire*, pp. 239–264.

Featherstone, M. (2000). *Body Modification.* London: SAGE Publications.

Fenske, M. (2007). *Tattoos in American Visual Culture.* New York: Palgrave Macmillan.

IFOP. (Juillet 2010). *Les Français et les tatouages.*

Haraway, D. (1990). A cyborg manifesto: Science, technology, and socialist-feminism in the 1980s. Dans *Feminism/Postmodernism.* New York: Routledge: Linda Nicholson.

Holz, C., Grossman, T., Fitzmaurice, G., & Agur, A. (May 5–10 2012). Implanted User Interfaces. *Proceedings of the SIGCHI Conference on Human Factors in Computing Systems (ACM CHI '12). Austin, TX, USA* (pp. 503–512). New York: ACM.

Jeannin, H. (2016). Quand les technosciences forment et informent: les nouvelles strates du corps. *Actes du colloque de Strasbourg, 28 au 30 avril 2016* (pp. 107–117). Nancy: PUN—Editions Universitaires de Lorraine.

Liotard, P. (2015). *Ceci est mon corps. Témoignages sur les modifications contemporaines de l'apparence.* Sous La Lime.

Marigny, J. (2008). Les métamorphoses du corps dans la littérature fantastique. Dans *Les imaginaires du corps en mutation.* Paris: L'Harmattan.

Mitchell, W. J. (2003). *Me+++: The Cyborg Self and the Networked City.* Cambridge, London: The MIT Press.

Steinberg, S. (2001). *La confusion des sexes. Le travestissement de la Renaissance à la Révolution.* Paris: Fayard.

Veyrat, N., Blanco, E., & Trompette, P. (2007/1 Vol. 1, n° 1). L'objet incorporé et la logique des situations. Les lunettes au fil de l'histoire et au gré des usages. *Revue d'anthropologie des connaissances*, pp. 59–83.

Vita-More, N. (2010). Epoch of plasticity: The metaverse as a vehicle for cognitive enhancement. *Metaverse Creativity, 1* (1), pp. 69–80.

Part IV

The Enlightenment Recovered

On Max More's Extropianism

Salomé Bour

Transhumanism is not something that can be easily defined. It is generally understood as a cultural and intellectual movement promoting human enhancement. But behind this movement lies a philosophical system that is less known: extropianism. This philosophy of transhumanism was set up by Max More in the 1990s. In order to understand the current forms of transhumanism, it is essential to know their roots and the principles on which they are based.

The Emergence of the Concept of Extropy

Extropianism was developed from the neologism "extropy". This term was coined by T.O. Morrow (Tom Bell), Max More's associate, in 1988 (January). It literally means "out of disorder". According to T.O. Morrow, extropy refers to a movement towards greater information.[1] But it was

S. Bour (✉)
rattaché au département de Philosophie, Centre d'Ethique Contemporaine, Université Paul-Valéry de Montpellier, Montpellier, France

P. Jorion (ed.), *Humanism and its Discontents*,
https://doi.org/10.1007/978-3-030-67004-7_9

Max More who gave to this metaphorical term its transhumanist definition. For Max More, extropy means "the extent of a system's intelligence, information, order, vitality, and capacity for improvement".[2] In other words, the term extropy is neither a scientific nor a technical concept. Rather, it expresses a dynamic whose objective would be to amplify progress in an unlimited way.

This concept emerges in a specific context. It must be related to the intellectual effervescence of California during the 1960s–1990s as well as the scientific interest in cryonics that prevailed during this period. Several cryogenics foundations were indeed created in 1964 and the first cryopreservation was carried out in 1970.

Nourished with science fiction as a child, Max More was particularly interested in life extension practices and increasing intelligence. In 1983, he was introduced to cryonics through a discussion group held at the Imperial College of London on Sunday morning. This group was composed of Luigi Warren, Mike Price, Garret Smyth and John de Rivaz. By mutual agreement, they had chosen to join and actively participate in Alcor Life Extension Foundation. At that time, the group's ambition was to create a cryogenics company in the United Kingdom to help promote immortalist ideas within the intellectual sphere. Alcor Foundation gained a new life in California in 1986, after leaving Arizona where it had been founded in 1972 by scientists Fred and Linda Chamberlain, based on the work of the physician and "father of cryonics" Robert Ettinger.

From 1984, Max More became a regular reader of Alcor's *Cryonics Magazine*, wherein he discovered the work of Eric Drexler, Mike Darwin, Robert Ettinger, Marvin Minsky and Ayn Rand's ethos, which would contribute to his already emerging philosophical project. After responding to a fundraising campaign of Alcor in January 1986, Max More got involved and met with Mike Darwin, the president of Alcor Foundation at that time, while he was completing studies in philosophy, politics and economics at Oxford University.

Aroused by Californian optimism, Max More left the United Kingdom after completing his degree to pursue research at the University of Southern California, where he got enrolled as a visiting student in 1987. It was between 1986 and 1988 that Max O'Connor (who only became known as Max More in 1990) met the student Tom Bell (T.O. Morrow) and developed the philosophy of extropy. The journal *Extropy* was created in 1988 as an outcome of their discussions and was dedicated to transhumanism.

Extropy has played a leading role in the development and dissemination of transhumanist philosophy. As the magazine's presentation indicates, the contributions cover several themes such as "life extension, immortalism and cryonics, artificial intelligence (AI), cognitive science and neuroscience, intelligence-increase technologies, nanotechnology, memetics, space colonisation, uploading, spontaneous orders (free markets, neural networks, evolutionary processes, etc.), science fiction, extropic psychology, futurist morality, reviews, transhumanism, and futurism". The diversity of these themes has aroused the interest of many philosophers and scientists who were to join the extropian and transhumanist discussion groups in the 1990s.

The philosophy of extropy was first presented by Max More in issue #1 of *Extropy*, published in the fall of 1988, of which there is no remnant copy. But issue #6, published in the summer of 1990, contains two texts key to understanding the nature of extropianism: the first, entitled "Transhumanism: Towards a Futurist Philosophy",[3] defines the philosophy of extropy as eupraxsophy, while the second outlines extropian principles, which were to be revised later several times.

Extropianism: An Eupraxsophy

The philosophy of extropianism is one of the atheist philosophies of life.[4] Paul Kurtz, an American humanist and sceptical philosopher, coined the term "eupraxsophy" in 1988. This neologism is a combination of three Greek terms—*eu* (good, well), *praxis* and *sophia* (wisdom, knowledge)—and literally means "good practice" or "active wisdom".[5]

When Paul Kurtz coined this term, he was working on the development of secular humanism. Secular humanism is a form of atheistic humanism that takes a rationalist and critical approach to dogmas, especially religious dogmas. It became popular during the 1980s. Kurtz's ambition was to reform traditional humanism in order to create a "New Enlightenment".[6] Also, eupraxsophy advocated emancipation from religion, which can only be achieved through a method of thinking that is both scientific and ethical and based on critical thinking. According to Paul Kurtz, the period of disillusionment and pessimism that followed World War II led to a renewed interest in religion. However, the consequence was a slowing down of scientific research in all fields. Thus, the "New Enlightenment" thought up by Paul Kurtz was aiming at kicking intellectuals out of their state of latency by allowing science to continue extending its field of knowledge and action and, by seeking to answer the question of the meaning of existence without any transcendental reference. It is in this perspective that Paul Kurtz expounded the values of eupraxsophy:

> "Eupraxsophy focuses on two main areas. First there are excellences intrinsic to the good life of the individual where freedom and autonomy, self-determination, and the right to privacy are respected, as well as the values of creativity, aesthetic appreciation, self-respect, self-control, and rationality. The ultimate goal is human happiness and joyful exuberance. Second are the principles of virtue and responsibility as they relate to other people in transaction communities. These include *the common moral decencies* of integrity, trustworthiness, benevolence, and fairness."[7]

These values are in conformity with Max More's principles and values of extropianism. Indeed, they share the same interest in individual freedom, autonomy and happiness. Moreover, the role assigned to virtue and responsibility in the societal context is also an essential tenet of extropianism. This role is particularly visible during Max More's Randian period, which focused on the "spontaneous social order" and its three essential values: trust, benevolence and justice. These values would continue to dominate Max More's period of Popperian inspiration, which was centred around the "open society". By defining extropianist philosophy as eupraxsophy, Max More's ambition was to embrace the values set out by Paul Kurtz.

In his essay, however, Max More distinguished extropianist philosophy from humanism, considered as the philosophy of existence that aims at developing the potentialities of the human being and giving value and meaning to human nature. The particularity of humanism is to value human nature and to give meaning to the finiteness of the human being. However, extropianism advocates the elevation of the human condition.[8] In other words, extropianism aims at going beyond humanism. This implies using technosciences to improve the human being in order to enable it to fully and perpetually express its potential. Unlike humanism, the philosophy of extropy rejects the idea that human progress is limited by its "nature". From an extropianist perspective, humanity must precisely overcome and improve its living conditions by artificially pushing back the limits of death, cognitive limits, emotional limits and so forth. For extropians, technosciences (NBIC), and mainly nanotechnologies of the early 1990s, could realise this ambition. But this project can only be realised if we accept detaching ourselves from the humanist conception of the human condition. Thus, the goal of extropianist philosophy is to bring about this transformation of mentalities by promoting the quest for immortality and the rational management of human evolution.

For Max More, however, religious beliefs are the main obstacle to the progress of mentalities. In continuity with Paul Kurtz, he positioned himself counter the religious model of an explanation of the world, phenomena as well as the meaning of existence. In his essay, he explains the persistence of religion by four causes: "Religion is (a) a pre-scientific system of explanation and technology; (b) a source of meaning, direction and emotional expression in life; (c) a means of social control; (d) a result of the structure of the brain in preconscious humans."[9] He reminds us to start that religion has a role of social control and domination, as Marx and Engels had laid bare. Max More argues that religion is an entropic force aiming at maintaining people in a state of alienation while preventing any form of progress. But he also indicates that there may be links between human cognitive functioning and the existence of religious beliefs. This controversial thesis of "bicamerality" had been developed by the American psychologist Julian Jaynes. According to Jaynes, the two hemispheres of the human brain held distinct functions before the emergence of consciousness: one represents the spirit of the gods (the right one), while the other (the left one) represents the spirit of man. Max More considers this thesis to be the most speculative cause. The third cause examined highlights the human need to understand the real and the existence of divinities. According to this theory, divinities are ways for providing meaning to what is not understandable, as well as to offer meaning to unusual or tragic phenomena.

For Max More, the religious explanation of the world could only exist in the absence of a scientific explanation of reality. He is also surprised by the persistence of religion when scientific knowledge progressed as well as growing empirical evidence of the ineffectiveness of religious beliefs. He claims that if religion still retains a place in contemporary society despite the answers provided by science, it is because human beings need in addition to give meaning to their existence. This is the fourth cause identified. For Max More, the problem is that religion maintains, through faith, the individual in a belief system that has been challenged by rational evidence. While the religious belief system offers spiritual immortality and bliss accessible after biological death, extropians want such goals to be achieved here on earth as soon as possible. To achieve this, it is necessary to begin by critically examining everything that could lead human progress to stagnate in order to counter it, while giving a meaning to human existence which would be based on reason.

In accordance with the concept of extropy, individuals must take charge of their future to make it better and they must not wait for the realisation of a divine plan that would exceed them. To encourage this, Max More wants to replace the pessimism and resignation imposed by religion with the extropian principle of "dynamic optimism" that invites the individual to develop his potential and transcend his faculties. "Forward, upward, and outward" is the motto of extropianism.

The emergence of extropianism is also to be put in perspective with different movements and the popularity of some of the theses that preceded it in the United States between 1960 and 1980: the hippie movement, the New Age movement, the theory of Intelligent Design, personal development and so on. Max More wants, in turn, to promote a philosophy of life. That philosophy aims at enabling everyone to develop and extend their potential infinitely. If religion proposes a spiritual immortality to which one reaches after biological death, extropianist philosophy proposes to make the quest for biological immortality a scientific objective.

But Max More will also include the extropian project in the continuity of Friedrich Nietzsche's thinking: "Religion justifies complacency and stagnation. The religionist has no answer to the extropian challenge put by Nietzsche's Zarathustra: 'I teach you the overman. Man is something that is to be overcome. What have you done to overcome him?'" If Nietzschean philosophy has indeed influenced extropian philosophy, the quote Max More uses here is particularly interesting. He makes Nietzsche's question the very statement of the extropian project: to ensure that we go beyond human, physical and psychological limits. Also, continuity with Nietzsche's philosophy is done on three levels: (1) criticism of religion as it prevents action; (2) an effort to counter nihilism that generates pessimism and a lack of values; and (3) an appreciation of the existence and transformation of the self in order to surpass the human being. But it should be noted that Max More reinterprets Nietzsche's ideas in order to serve his own extropian project.

In short, Max More seeks to fight an anti-progressive thought that tends to unfold when humans have the power to transcend their own limits. The principles of its philosophy must be sufficiently attractive to dislodge religious, stagnant and pessimistic thinking. The goal of extropianism is then to provide a set of principles and values that give meaning to individual existence and an orientation to society to foster progress. Extropianism is a fundamentally optimistic philosophy, a "vaccine for the future shock" to amplify the rate of change of people's mindsets in order to develop extropy, guided by science and reason.

Extropian Principles

The first version of the extropian principles was published in 1990 (summer) and was constantly updated by Max More until 2003. They are based on four values: critical thinking, creative or innovative thinking, collaboration and communication. According to Max More,

> "The Principles of Extropy do not specify beliefs, technologies, or policies [...]. The Principles of Extropy do consist of a handful of principles (or values or perspectives) that codify proactive, life-affirming and life-promoting ideals. [...] The Principles are intended to be enduring, underlying ideals and standards. At the same time, both in content and by being revised, the Principles do not claim to be eternal truths or certain truths."

The primary objective of the extropian principles is to promote the transition from stagnant thinking to proactive thinking oriented towards progress and the improvement of humankind's living conditions. Extropian principles derive from the concept of extropy and form the philosophical basis of transhumanism. It is based on these principles that the 1998 *Transhumanist Declaration* will be drafted. In their most recent version (2003), there are seven of them: perpetual progress, self-transformation, practical optimism, intelligent technology, open society, self-direction and rational thinking.

Perpetual progress, initially "unlimited expansion", values any action whose objective is the search for more intelligence, efficiency and longevity. This principle invites us to see the obstacles that could arise in the face of extropians as challenges. To maintain the momentum of progress, it is up to extropians to elevate the human condition by transcending natural boundaries. The development of humanity's potential can only be achieved through the constant search for perfection, with the aim of continuous improvement.

However, this continuous improvement can only be achieved by seeking to transform oneself. This transformation is first inner and involves an examination of one's own beliefs and certainties that could prevent one's own evolution towards perfection. Therefore, it requires a work on oneself before it can be extended by a physical transformation. Indeed, physical transformation is not just aesthetic, it must be the result of the identification of the *true self* that technosciences will make possible by increasing, for example, the cognitive or emotional capacities of the individual. In short, it is a question of perpetual physical, intellectual, psychological and ethical improvement. The principle of self-transformation implicitly introduces the concept of morphological freedom, that is, the right of every individual to freely modify his body, using technosciences. From an extropianist perspective, the freedom to choose one's physical form does not depend on the individual's desires, but on a rational choice that allows the expression of the individual's "true" personality. Proponents of extropianism are described by Max More as rational, autonomous people whose projects succeed because they are driven by reason.

In addition, extropians adopt the principle of "practical optimism" and dynamic optimism, the objective of which is to promote proactive thinking. Indeed, rather than waiting and dreaming of a better world, or hoping for a life after death, extropians have the ambition to create their own future. In other words, everyone can innovate, everyone can act so that what is ideal can become real. It means adopting an optimistic view of progress and not giving in to passivity, pessimism or defeatism. Extropian optimism is "realistic" because it takes into consideration all the data to make an objective choice in relation to the projects to be carried out. For example, rather than considering the phenomenon of senescence as inevitable, extropians refuse to accept this phenomenon and will do everything possible to fight it. Scientific results show that it is possible to delay ageing and that is why this project is seen by extropians as rational. The use of reason implies a realistic examination of what is possible or not and what is desirable or not. Decision-making must also be accompanied by ethical reflection, which must not constitute an obstacle to progress.

Progress can only result from the intelligent use of technology, as stated in the fourth principle (intelligent technology). Extropianism values the use of technology to transcend human nature by ridding it of defects such as disease, ageing, death or negative emotions (sadness, jealousy, fear and anxiety) and behaviours that are harmful to oneself (stress and disorders) and others (aggressiveness and violence). The application of science and technology must then be oriented towards an end that is presented as "good" for humans, which necessarily implies an ethical use of the latter. But an intelligent use of technologies also involves accompanying the physical and psychological transformation of extropians.

However, the unlimited pursuit of progress can only be achieved in a favourable socio-political and economic context. The fifth principle of "open society—information and democracy" is a 1993 addition and was first called "spontaneous order". Originally, this principle was connected to the philosophy of Ayn Rand, who influenced Max More in his early days. It was at the time when he developed extropian epistemology in 1994 that he broke away from Ayn Rand's objectivism and became interested in the notion of an open society resulting from Karl Popper's work.

The extropian open society model is a model of social and political organisation whose objective is to promote the dynamics of the extropy phenomenon. There exist values specific to democracy, at the same time there exist also elements of political and economic liberalism. The political system favoured by extropianism in 2003 was liberal democracy. In contrast with the libertarian conception that the state must be ultra-minimal and with Ayn Rand's utopian system that does not include the idea of a state, liberal democracy implies a measured presence of the state. The latter cannot be authoritarian or coercive. Its objective is to structure the company. It is aimed towards an end: happiness and progress. It also guarantees individual freedom as defined by extropians.

The goal of an open society is to promote the constant transformation of society towards the best, that is, greater happiness and freedom, an emancipation from all the determinisms linked to nature and the human condition. But this quest for perfection is not a utopia. The utopia of the enlightenments is the representation of a society as the best possible. However, for Max More, the extropian project is to pursue an "extropia", that is, a society that is not static. Extropia is based on the idea that the full actualisation of human potential is never achieved. It also implies a constant review of its own model and the progress made. This socio-political model promotes the development of extropy without any barriers, whether natural or cultural. In an open society, individual freedom is preserved: the individual acts rationally and autonomously, making use of his critical and creative mind.

The principle of self-direction emphasises it too. This principle consists in taking charge of oneself, being autonomous and making personal choices, that is, not oriented by societal or political norms:

> "Each individual should be free and responsible for deciding for oneself in what ways to change or to stay the same. Self-direction means being clear about our values and our purposes. Having clear purpose in life not only brings both practical and emotional rewards but also protects against manipulation and control by others. Freedom from others brings fulfilment and personal progress only when combined with self-direction."

In other words, the philosophy of extropy puts the individual at the centre, as master of himself, as being able to use his own understanding according to the values he has chosen rationally and according to his own objectives. Also, an open society requires rational individualism, or "cognitive independence" as Max More call it. The individual who seeks to realise his or her potential must do so freely while being aware of the risks he or she takes when carrying out experiments that could promote his or her improvement.

Responsibility and freedom are at the heart of the individual's conception of the individual within the political framework of the open society. Indeed, although the open society favours the individual, the collective must not be forgotten. That is why benevolence is also necessary. The development of the individual is based on his relationships with others, which must be respected in their choices. In an open society, everyone is free to think, act and experiment according to their personal interests, while accepting responsibility for themselves because no one forces them to act. This freedom can only spontaneously promote collective prosperity if it acts in a benevolent way. Everyone's objective would then be to maximise profits and minimise negative effects. Thus, for Max More, it is in the common interest to go beyond biological, physical and social limits to live forever, to become happier and to extend the potential of humanity.

But self-orientation must be part of the reason and must emancipate itself from beliefs, dogmas and prejudices. This is stated by the principle of "rational thinking". This principle emphasises the development of critical thinking to avoid anything that could stagnate personal development, progress and innovation. Just as the scientist constantly seeks to test the theories he builds, if he never produces eternal truths, so the extropian takes a critical approach by continually questioning his ideas. In Max More's view, this attitude is a strictly extropian attitude: it puts reason at the centre of all thought and action.

The extropian individual is optimistic and can emancipate himself from collective thinking, to resist disenchantment, defeatism. For example, if we cannot fight ageing today, even if discoveries suggest that this will be possible within a variable number of decades, the extropian would not accept the common defeatist conception by thinking that "this is impossible" or by saying: "we will never succeed". On the contrary, the proponent of the philosophy of extropy would rather see the scenario as follows:

1. Objectively, ageing causes more suffering and damage than benefits. According to this utilitarian calculation:
2. There is an ethical imperative to fight ageing: there are more individual and social benefits to annihilating this phenomenon than to preserving it.
3. Reason therefore requires us to fight against this phenomenon: it is then necessary to understand its mechanisms in order to be able to act. On the one hand, some living beings are not subject to ageing, and on the other hand, new knowledge is emerging through studies and suggests the possibility of delaying the ageing phenomenon in certain animals such as mice (SENS Foundation).
4. Therefore, it is rational to think that ageing could be eradicated for human beings if research continues to produce results in this direction.

The scientific method consists in taking any knowledge for an indisputable truth, otherwise it would be like adopting a dogmatic position. This method also requires us to constantly question any knowledge, to maintain a momentum. For extropians, reason can decode reality with the help of scientific tools like critical and creative thinking, Ockham's razor, Popper's concept of falsifiability, experimental scientific method and so on. Then, Max More will also develop an extropian epistemology, initially inspired by Ayn Rand's objectivism.

In short, the seven principles of extropianism are a manifesto. It proposes both an ideological system centred around the idea of perpetual progress based on the concept of extropy and a set of values promoting an attitude based on reason. The *Extropy Institute* has played a considerable role in the dissemination of extropian ideas with its mailing list, created in 1991 by Harry Hawk and Perry Metzger. This "incubator" of new ideas was used to create the foundations of the philosophy from which transhumanism would develop. Indeed, it formed the first true transhumanist communication network where extropianism was discussed by a group of researchers from the philosophical, scientific and technological fields. It is also from this network that the *World Transhumanist Association* would be founded in 1998 by philosophers Nick Bostrom and David Pearce.

Pancritical Rationalism and Extropian Epistemology

Obviously, the philosopher Ayn Rand was initially a major influence on Max More's thinking. Indeed, Max More's interest in rationality and spontaneous order is largely rooted in the work of the founder of objectivism. However, Max More would move away from it by developing his own epistemological system in accordance with extropianism. It was the American computer scientist Mark Miller who pointed out, in an interview published in 1993 in issue #11 of *Extropy*, that extropianism lacked true epistemological questioning, while recommending Max More that he reads *The Retreat to Commitment* by William Bartley, a philosopher specialised in Popperian epistemology. The Randian rationalist epistemology was then replaced by the concept of "pancritical rationalism" developed by William Bartley, which was more in line with the philosophy of extropy.

In a conference given during the first conference cycle EXTRO-1 organised by the Extropy Institute in 1994, Max More presented his epistemological work. Through this paper entitled "Pancritical Rationalism: An Extropic Metacontext for Memetic Progress", Max More attempts to show how "pancritical rationalism" fits with extropian principles.

As he explained, Ayn Rand's system tends to become dogmatic. For Max More, the major problem is the "Foundationalist nature" of the epistemological part of Ayn Rand's philosophy. This nature implies that this system cannot be revised, that it must be accepted without being called into question. The Randian system is therefore posited as a truth:

> "Foundationalism shows up first in the axiomatic foundations of Objectivism. Rand declared the ideas of existence, identity, and consciousness to be axiomatic concepts. […] So, we begin with unquestionable axiomatic concepts like existence, identity, and consciousness […] that cannot rationally be questioned, then we straightforwardly derive all knowledge from these foundations by means of the guaranteed validity of our senses. The truth is manifest, to Rand, and anyone failing to reach her conclusions (assuming they have the intellectual capacity to follow the reasoning) must be deliberately sabotaging their mind, they must be irrational and therefore evil."

However, extropianism rejects the existence of stable and indisputable axiomatic foundations and fights all forms of dogmatism, emphasising the constant revision of its principles and values. The difficulty that More faces is the need to constantly reassess the foundations. This imperative could lead, for example, to infinite regress and scepticism. Similarly, one may wonder how to base knowledge without falling into a form of dogmatism.

Max More first read the sceptical epoch and the epistemological approach of the sceptics as well as David Hume's scepticism to show that while they constitute bases, they cannot be the basis for extropianist epistemology. But Popper and Bartley's readings will open Max More to the notion of "epistemological fallibility".

However, Popper's critical rationalism does not satisfy Max More because it ends up promoting an "irrational faith in reason". If Popper does not advocate irrationality, he nevertheless criticises non-critical rationalism and excessive rationalism ("comprehensive rationalism", "self-contained") by proposing a critical rationalism capable of limiting its degree of irrationality as much as possible. According to Popper, acceptance of the principle of rationality depends on a moral choice of the individual, in the sense that the decision may have consequences for society. Therefore, the position of critical rationalist, even if it may include a part of irrationality—which must be as minimal as possible—remains preferable to total irrationality, whose effects are generally negative because they depend on emotions and passions. As Max More explained, Popper proposed in his essay *The Open Society and Its Enemies* to adopt "a minimal concession to irrationalism" to avoid infinite regress:

Whoever adopts the rationalist attitude does so because he has adopted, without reasoning, some proposal or decision, or belief, or habit, or behaviour, which therefore in its turn must be called irrational. Whatever it may be, we can describe it as an irrational *faith in reason*.... *t*he fundamental rationalist attitude is based upon an irrational decision, or upon faith in reason.

However, what disturbs Max More is the idea of accepting this part of irrationality, and the idea that we should accept a "principle of rationality" in which the critical rationalist person would have "faith": these are two elements that his principles seek to combat.

Bartley's "pancritical rationalism" then appeared to be the solution to the Popperian problem. To understand Bartley's contribution to Popper's critical rationalism, it is necessary to recall the distinction established by Popper: he distinguishes non-critical rationalism and comprehensive rationalism from critical rationalism, showing that the second approach is morally preferable to the first one that leads to *aporia*. However, the second approach necessarily leads to the acceptance of a principle of rationality that is unfounded and impossible to establish. Bartley, therefore, considered a revision of critical rationalism by assuming that everything can be subjected to criticism. This prevents from the adoption of absolute foundations. In other words, for Bartley, no irrational basis can be tolerated, as he states in *The Retreat to Commitment*.

Pancritical rationalism takes up the traditional sceptical approach while being part of the modern scientific approach which affirms that any postulate must be able to remain open to criticism. Bartley established also a distinction between criticism and justification. For him, propositions must be criticised rather than justified. He claims that everything must be criticised, without any attachment to any presupposition or belief, which prevents from a dogmatic attitude. Bartley follows Popper in the debate on the rationality of science that emerged from the 1960s onwards, in relation to the critique of Anglo-Saxon logical positivism, and which will lead to a form of "rationality crisis", as the philosopher Sandra Laugier said.

For Max More, the constant re-examination of assumptions, beliefs or knowledge is fundamental in the extropian process. He also points out that rationality is not synonymous with omniscience: it is possible to have beliefs, but it is the attitude towards these beliefs that is decisive.

Then, pancritical rationalism complements well extropian principles and more specifically the principle of rational thought. However, Max More points out that this is not a component of the philosophy of extropy, but rather a meta-context of the extropian context. In other words, pancritical rationalism is a rationalist position that accompanies extropians in their process of self-reformation: it could help to rid them of a kind of fear of criticism that prevents individuals from moving towards progress. Pancritical rationalism is like a pretext for "cognitive strategies". These strategies help to foster dynamic optimism and to be more open to constructive criticism. The extropian project and its epistemology must support the transformation of mentalities. The social and political application of Karl Popper and William Warren Bartley's work is then interesting for Max More. However, it should be noted that Karl Popper has links with the philosopher and economist Friedrich Hayek, a thinker of the "Great Society" and an important figure in liberalism, and that William Warren Bartley is both the biographer of Popper and the likely true author of *The Fatal Conceit. The Errors of Socialism* (1988), a book attributed to Hayek, supposedly written by him at a time he was critically ill. Similarly, Ayn Rand's work has also been instrumental in shaping contemporary American political thought and has often been put into perspective with Friedrich Hayek's work. The Popper/Hayek/Rand trio has influenced Max More in the development of his political and economic thinking. William Bartley's reading completes this panel and allows Max More to open his thinking to the "open society" model proposed by Karl Popper, towards which he is moving. It marks the moment from which he partially abandoned *laissez-faire* capitalism in favour of a political conception of orientation towards liberal democracy.

But that is not enough to say that Max More is a libertarian. Indeed, in accordance with extropian principles, it does not promote any political system. It should be noted that Max More does not seek to promote any specific policy. Despite this, it does not mean that extropianism accepts any form of political model. The liberal or even libertarian branch is more suitable for extropianism because of the predominant place of freedom within extropian principles. Max More discovered libertarianism very early on at seventeen through Murray N. Rothbard's book *For a New Liberty: The Libertarian Manifesto*. Murray N. Rothbard was a philosopher and economist, member of the Vienna School as Friedrich Hayek, whose work is mentioned by Rothbard. This is the source of Max More's interest in the spontaneous order developed by Friedrich Hayek, a principle he developed and put into perspective with Ayn Rand's ideas.

Thus, the development of extropianism occurred in two steps: first a libertarian orientation step and then a step oriented towards liberalism. This change resulted from Max More's readings but also from the exchanges via the mailing list of the *Extropy Institute*. It must be added the creation of the *World Transhumanist Association* (now *Humanity+*)

which developed at the same period. The wider dissemination of transhumanist ideas led Max More to nuance his political position, particularly in relation to the ethical issues that will emerge at that time.

It must be remembered from the latter's epistemological choice the willingness to propose an approach that makes it possible to value critical and creative thought while preventing oneself from any dogma. Extropianism is above all a philosophy of action. Therefore, in the continuity of his epistemology, Max More elaborated around 2004–2005 the *Proactionary Principle*, whose objective is to lead individuals to reconsider their relationship to technical creations and the perspectives they open by placing the emphasis on action and not on prevention. This principle is designed as an alternative to the precautionary principle accused of slowing down innovation. If ethical reflection already underpinned the philosophy of extropy, it was at this point that Max More formalised it. One reason for this is that the spread of transhumanism at the international level has generated fears.

Conclusion

If the place of ethical reflection in extropianism and in the transhumanist movement deserves further examination, the objective of this chapter was first and foremost to review the philosophical foundations of transhumanism. Indeed, if transhumanist ideas were deployed from 1998 onwards, they were first set out through Max More's work. The concept of extropy, the extropian principles and the extropian epistemology form the basis for reflection from which the movement has been able to build and strengthen itself. At the beginning of the transhumanist movement, the Extropy Institute played a decisive role by bringing together researchers interested in transhumanist ideas and by making Max More's philosophy known through its mailing list. Extropian philosophy could only evolve thanks to it. The reductive view of extropianism such as this philosophy was created by Max More prevents his influence in the transhumanist movement from being sufficiently considered. However, it is based on its examination that transhumanist philosophy can be truly understood and criticised.

Notes

1. T.O. Morrow's message in the extropian mailing list. Retrieved April 30, 2019, from https://extropians.weidai.com/extropians.4Q97/0985.html.
2. Max More's message in the extropian mailing list. Retrieved April 30, 2019, from http://extropians.weidai.com/extropians.96/4101.html.
3. More, Max. Transhumanism: Towards a Futurist Philosophy. *Extropy*, 1990, vol. 6 (Summer), pp. 6–12.
4. More, Max. The Philosophy of Transhumanism. In: More, Max and Natasha Vita-More (ed.), *The Transhumanist Reader: Classical and Contemporary Essays on the Science, Technology, and Philosophy of the Human Future*. Chichester, West Sussex, UK: John Wiley & Sons, 2013, pp. 3–17.
5. More, Max. Transhumanism: Towards a Futurist Philosophy. *Extropy*, 1990, vol. 6 (Summer), p. 8.
6. Kurtz, Paul. The Re-enchantment: The New Enlightenment. *Free Inquiry Magazine*, vol. 24, n° 3, April/May 2004.
7. "Introduction: Up from Atheism." In: BUPP, Nathan. *Meaning and Value in a Secular Age: Why Eupraxsophy Matters—The Writings of Paul Kurtz*. Amherst, NY: Prometheus Books, 2012.
8. More, Max. A Letter to Mother Nature. In: More, Max and Natasha Vita-More (ed.), *op. cit.*, pp. 449–450.
9. More, Max. Transhumanism: Towards a Futurist Philosophy. *Extropy*, 1990, vol. 6 (Summer), pp. 6–12, p. 6.

References

Bartley, William W. III. *Retreat to Commitment*. London: Open Court, 1999.

Bupp, Nathan. *Meaning and Value in a Secular Age: Why Eupraxsophy Matters – The Writings of Paul Kurtz*. Amherst, NY: Prometheus Books, 2012.

Kurtz, Paul. "The Re-enchantment: The New Enlightenment". *Free Inquiry Magazine*, vol. 24, n° 3, April/May 2004.

Laugier, Sandra. "De la logique de la science aux révolutions scientifiques". In: Wagner, Pierre, and Jocelyn BENOIST (ed.). *Les philosophes et la science*. Paris: Gallimard, 2002, p. 964–1016.

More, Max. "A Letter to Mother Nature." In: More, Max and Natasha VITA-MORE (ed.), *The Transhumanist Reader: Classical and Contemporary Essays on the Science, Technology, and Philosophy of the Human Future*. Chichester, West Sussex, UK: John Wiley & Sons, 2013a, p. 449–450.

More, Max. "Extropian Principles", v. 3.11, 2003. Retrieved April 30, 2019, from https://web.archive.org/web/20131015142449/http://extropy.org/principles.htm.

More, Max. "Pancritical Rationalism: An Extropic Metacontext for Memetic Progress", 1994. Retrieved April 30, 2019, from https://web.archive.org/web/20180117163844/http://www.maxmore.com:80/pcr.htm.

More, Max. "The Extropian Principles". Extropy, 1990a, vol. 6 (Summer), p. 17–18.

More, Max. "The Extropian Principles", v. 2.5, 1993a. Retrieved April 30, 2019, from http://www.aleph.se/Trans/Cultural/Philosophy/princip.html.

More, Max. "The Philosophy of Transhumanism". In: MORE, Max and Natasha Vita-More (ed.), *The Transhumanist Reader: Classical and Contemporary Essays on the Science, Technology, and Philosophy of the Human Future*. Chichester, West Sussex, UK: John Wiley & Sons, 2013b, p. 3–17.

More, Max. "Technological Self-Transformation: Expanding Personal Extropy". *Extropy*, 1993b, vol. 10 (Winter/Spring), p. 15–24.

More, Max. Transhumanism: Towards a Futurist Philosophy. *Extropy*, 1990b, vol. 6 (Summer), p. 6–12.

Popper, Karl. *The Open Society and Its Enemies*. London: Routledge, 2011

Younkins, "Edward W. Ayn Rand and Friedrich A. Hayek: A Comparison". *Libertarian Papers*, 2017, vol. 9, p. 84–113.

World Transhumanist Association. "The Transhumanist Declaration" v. 2.4 (1998). Retrieved April 30, 2019, from https://web.archive.org/web/19980702105748/http://www.transhumanism.com/declaration.htm.

Transhumanism and Advanced Capitalism: Elitist Logics and Dangerous Implications

Alexander Thomas

The tenor of much academic bioethical thought dedicated to the potentially profound changes radical technological development may bring about is frequently characterised by abstraction. Often this is because the possible advancements NBIC technologies (nanotechnology, biotechnology, information technology and cognitive science) may bring are deemed so radical as to render the prevailing social conditions irrelevant. However, the nature and pace of the changes do not exist in a social vacuum. The prevailing societal structures will play a significant role in the process and its effects. Transhumanists often overplay the instrumental nature of technology and fail to thoroughly situate technological advancement within the social, economic, cultural and political context from which it is emerging. Max More and Natasha Vita-More claim: 'One aspect of transhumanism… is the need for inclusivity, plurality, and continuous questioning of our knowledge' (2013: 1). It is my contention that these

A. Thomas (✉)
London, UK
e-mail: a.thomas@uel.ac.uk

P. Jorion (ed.), *Humanism and its Discontents*,
https://doi.org/10.1007/978-3-030-67004-7_10

three principles are incompatible with transhumanist development in the context of an advanced capitalist logic. This chapter focuses on the principle of inclusivity.

Implications and Trajectories of Advanced Capitalism: Inequality, Expulsions and Concentrations

Perhaps the single most significant trend that has arisen during the most recent incarnation of global capitalism is a growth in inequality within developed Western countries, and between the wealthiest 1% and the rest globally. These dynamics undermine the idea that a process such as the development of transhumanist technologies will be an inclusive one. Between 1980 and 2014 the average real income of the richest 1% of Americans increased by 169% from $469,403 to $1,260,508 (inflation adjusted) (Stiglitz, in Jacobs and Mazzucato 2016: 135). This equates to a rise in the share of national income from 10% to 21%. The inequity intensifies further up the wealth scale. As Dorling reveals 'the 1% have pulled away, inequalities within the 1% have grown enormously' (2014: 11). It is no surprise then that over the same period, the top 0.1% saw their average real income increased by 281% from $1,597,080, to $6,087,113 (inflation adjusted) meaning 'their share of national income almost tripled, from 3.4 to 10.3%' (Stiglitz, in Jacobs and Mazzucato 2016: 135). Despite the economic growth in the USA between 1983 and 2009, the bottom 80% of the income distribution saw a net fall in their wealth over this period (Wolff and Allegretto, in Brynjolfsson and McAfee 2014) completely undermining the veracity of trickle-down economics. Increases in inequality measured by wealth are even more pronounced than those measured by income, 'by one estimate more than 10 times so' (Stiglitz, in Jacobs and Mazzucato 2016: 136). It is estimated that 26 individuals now have the same amount of wealth as half of the world put together (Oxfam 2019), and the $550 billion added to the wealth of billionaires in 2017 alone would be enough to wipe out global poverty seven times over (Oxfam 2018).

Whilst the impact of technology on the economy has exacerbated this inequality, technology is not the sole cause—elite groups have vested interests in exploiting existing power structures. Technology is an additional enabler that is more accessible to those with the power, money and know-how to best leverage it. As Saskia Sassen argues,

> Rich individuals and global firms by themselves could not have achieved such extreme concentration of the world's wealth. They need what we might think of as systemic help: a complex interaction of these actors with systems regeared toward enabling extreme concentration. Such systemic capacities are a variable mix of technical, market, and financial innovations plus government enablement. (Sassen 2014: 13)

Meanwhile, at the lower end of the social pyramid is an 'emergence of new logics of expulsion' (Sassen 2014: 4). The combination of concentration and expulsion is effective in 'capturing the pathologies of today's global capitalism' (Sassen 2014: 4). Once more the nebulous systemic logics play a key role in enabling the process and determining its opacity creating a centrelessness which is integral to the resilience of the system.

> Historically, the oppressed have often risen against their masters. But today the oppressed have mostly been expelled and survive a great distance from their oppressors. Further, the 'oppressor' is increasingly a complex system that combines persons, networks, and machines with no obvious centre. (Sassen 2014: 10)

Surplus populations of expelled people serve as a useful tool for capitalist interests. They suppress wages by offering a large reserve pool to call upon during periods of growth and can also be used as a disciplinary measure or threat during periods of stagnation. As well as suppressing wages, they weaken workers' bargaining power which can lead to worsening employment conditions. In modern advanced capitalist economies, there is an increase in 'precarity' characterised by 'more casual working hours, low and stagnant wages, decreasing job protections and widespread insecurity' (Srnicek and Williams 2015: 93). Sassen (2014) observes that this precarity accentuates rises in depression, anxiety and suicide.

The treatment of surplus populations has implications in a world with transhuman technological potentialities. More than 60,000 migrants have lost their lives on fatal journeys in the last 20 years (IOM 2017), and 120,000 deaths in the UK between 2010 and 2017 were linked to health and social care cuts (BMJ 2017). These deaths can be said to result from systematic marginalisation. As labour tends to play a subordinate role in an info-tech economy we may see a significant rise in these expulsions. Numerous commentators (e.g. Ford 2015; Lanier 2013; Srnicek 2017) point out 'platform-capitalist' companies such as Uber, Airbnb and Netflix require very few staff to run their multibillion-dollar businesses. Indeed, the pattern is ubiquitous: 'industries which today are nascent and fast growing, and tomorrow will be economic giants, are extremely parsimonious employers of humans' (Chace 2016: 28). Furthermore, the rise of artificial intelligence and robotics facilitates the automation of the workplace in a wide variety of fields. Some capitalists herald automation as the business world's 'emancipation from human workers' (Little, in Carr 2015: 37). Thus humans will not be set free from the drudgery of work, rather business will be set free from human fallibility and co-dependency.

Whilst it would be overly deterministic to suggest that mass automation unemployment is inevitable, it is important to beware of the possibility. If changes occur at a pace consistent with a technological ontology rather than a human one, then the collateral damage could be significant. If robots and AI do render masses of people redundant 'the most important question in 21st-century economics may well be: What should we do with all the superfluous people?' (Harari 2017). A small elite could gain an almost total concentration of wealth: an elite with access to the most powerfully transformative technologies in world history and a redundant mass of people no longer suited to the evolutionary environment in which they find themselves who would be entirely dependent on the benevolence of that elite. As we have seen with the example of migrants, prevailing liberal values do not always extend to those who do not share the same privilege, race, culture or religion. But even as far as they do extend, there has been a pragmatic reason for these liberal values to develop: 'Liberalism succeeded because there was much political, economic and military sense in ascribing value to every human being...There was value to every pair of hands that could hold a rifle or pull a lever' (Harari 2016: 308). History would suggest it is naïve to think, given the likely difference in cultural worlds and even potential species-defining variations between the elite and the rest, that such altruism would be forthcoming. Certainly, if this era of advanced capitalism is the period which ushers in such radical technological change, then the logics of expulsions (and the normalising force of this pattern) have deeply disturbing cultural implications. In an era of radical technological power, the 'superfluous' masses may present a significant security threat to the elite, which could be used to justify aggressive and authoritarian actions.

Savulescu: Saving Humanity with Enhanced Morality

Alongside these economic logics and processes, transhumanist conceptions of humanity are often explicitly misanthropic, adding to concerns about the potential for advanced capitalist logics and transhumanist aspirations to have dehumanising effects. Humankind can be viewed as desperately in need of salvation from the myriad dangers that pose an existential threat at the beginning of the twenty-first century. Technology, especially of the sort that intervenes in our very nature, is increasingly perceived as that salvation. For example, Minsky argues that 'we are unlikely to last very long—on either cosmic or human scales of time. In the next hundred or thousand years, we are liable to destroy ourselves, yet we alone are responsible not only for our species' survival but for the continuation of intelligence on this planet and quite possibly in the universe'. (Cited in Garreau 2005: 123)

Julian Savulescu (2009) also cites species survival as a primary justification for human enhancement, arguing that we face a 'Bermuda Triangle of Extinction': radical technological power, liberal democracy and human moral nature, the triumvirate which threatens a potential cocktail of destruction. Fuller and Lipinska characterise the thinking of Nick Bostrom and Savulescu thus:

> The various diseases of affluence from which humanity suffers today—from ecological degradation to mediocre health maintenance and tolerance of massive socioeconomic disparities—are traceable to a failure of our biological hardware to catch up with the aspirational software (i.e. the progressive philosophies, sciences, etc.) that successive generations of humans have programmed into the hardware. (2014: 84)

Their description is instructive; firstly, in its dehumanising, mechanistic language (perceiving humanity as something akin to a bug-ridden computer system), but also because it places morality as the particularly faulty aspect of human nature.

Conceptions of 'human nature' are highly contentious, but it is in this area that Savulescu (2009) believes we have the greatest capacity to alter the alarming future he foresees. For Savulescu, the causes of problems are to be found within our biological wiring. There is no attempt to explain the complex systemic social logics or cultural dynamics that play a significant role in their development. In his view, the integral failings at the moral core of human nature require technological intervention to prevent us from self-destruction. For Savulescu there is an 'urgent imperative to enhance the moral character of humanity' (Persson and Savulescu 2008: 1) exactly because radical technological potentialities are becoming possible. We cannot wait for natural evolution to catch up with the shifting world we find ourselves in: it is too slow and lacks the requisite insight.

Our moral dispositions and choices create social inequality. One way to address that inequality is to change the dispositions of people who make those choices. After all, gross social inequality already exists. It is time to treat the cause rather than the symptom of the disease. And the disease is our limited set of moral dispositions (Savulescu 2012: 196).

Savulescu simplifies all social inequality as being caused by human dispositions. But this ignores the fact that our moral dispositions themselves derive from social contexts. If 'moral dispositions' here refer to the biological wiring of all human beings, then it is a dubious claim as it ignores the social contingency of each individual morality, the variance of 'moral dispositions', the divergent positions of power people wield. Elsewhere, he is more cautious: 'embracing biological enhancements does not imply biological causation. There is nothing intrinsically wrong with employing biological solutions to social problems' (Savulescu 2012: 192). However, a lack of awareness of the roots of social problems may well result in biological solutions that exacerbate rather than ameliorate such issues. Social and biological reality are intricately connected and co-evolved. Whilst the failings of humankind to deal with global problems are increasingly manifest, deep social as well as biological understandings are required. Moral dispositions are neither self-constituted (merely biological) nor all determining. Savulescu's insistence on divorcing biological and social systems in his analysis underplays the complexity and inter-related nature of these complementary systems. Indeed systemic logics that determine vested interests and cultural norms play a significant role in how moral dispositions are shaped. Savulescu fails to situate our moral failings within the wider cultural, political and economic context.

Secondly, it is necessary to consider how these technologies will be disseminated, prescribed and potentially enforced in order to address the moral failings they seek to 'cure'. Whilst it is easy to imagine certain broad mental enhancements, such as intelligence, would be desirable to people, the idea that 'moral' enhancements would appeal is more dubious. Setting aside for a moment the difficulty in determining exactly what we may mean by 'enhancing moral nature', it is clear that being more empathetic or altruistic may be a distinct disadvantage in the context of a competitive capitalist society. Many of the altruistic attributes that Savulescu wishes more people would exhibit would render them less attuned to a fitness landscape where a mantra is 'greed is good'. Indeed, this is one of the reasons why arguing that human moral nature is bad per se is contestable: the advanced capitalist system encourages certain morally dubious behaviours. In fact human moral nature is extremely pliable and varied. Responsibility for the dissemination of moral enhancements would likely reside in the current power structures which may well bear much of the responsibility for these failings in the first place. It is important therefore to thoroughly interrogate the structures and tendencies that exist within our political and economic framework in order to identify the inherent biases within the system. It is these biases that will determine the development, design, accessibility and utilisation of technologies that Savulescu cites as offering strong possibilities for altering our failing moral nature.

Finally, the conception of 'morality' may differ widely between groups. The logics of advanced capitalism play a role in determining the culturally accepted norms of morality also. In Savulescu's work, although he speaks of promoting social justice and 'good' there is much to suggest his view of morality is a fairly socially conservative one. It is easy to conceive of various moral dangers that radical technological enhancement will create in the future, but Savulescu tellingly focuses his attention on individual actors: 'in the next decades there's no reason why a backyard terrorist or fanatic or psychopath could not create a bio weapon that would at least destroy millions of lives' (Savulescu 2009). While such a reality is imaginable, it is worth noting the prominent position that terrorism plays in the public imagination in developed Western countries today and its relationship to the dehumanising treatment of migrants. Furthermore, it plays an important role in popularising certain policies that tend to cede civil liberties. Indeed, despite advocating 'fixing' human moral nature, Savulescu claims that the technologically enhanced threat from terrorists and other malign individuals requires the relinquishing of civil liberties including the right to privacy:

> … we will need to relax our commitment to maximum protection of privacy, we're seeing an increase in the surveillance of individuals and that will be necessary if we are to avert the threats that those with anti-social personality disorder, fanaticism, represent through their access to radically enhanced technology. (2009)

Savulescu's argument illustrates Debra Shaw's concern that 'acquiescence to increased surveillance and regulation of the Internet is secured on the basis of the anxieties mobilised by the threat of terrorism, concerns about copyright theft and child protection issues' (2008: 37). In fact this surveillance allows corporations and governments to access and make use of extremely valuable information. The subterranean ideological reasons for surveillance tie into the technocapitalist system's logic as it has transitioned to an information economy: 'the ultimate commodity in this post-Fordist economic structure is information' (Shaw 2008: 27). Lanier explains: 'troves of dossiers on the private lives...of ordinary people, collected over digital networks, are packaged into a new private form of elite money...It is a new kind of security the rich trade in... It becomes a giant scale levee inaccessible to ordinary people' (Lanier 2013: 99). This levee is not only inaccessible but also imperceptible. Its impact extends beyond skewing the economic system towards elites with access to 'siren servers' (Lanier 2013) but also results in significant curtailments of liberties as power is both radically more effective and nebulous. It is telling that whilst Savulescu proposes to fix social problems with technological solutions, he simultaneously endorses the loss of rights for individuals on account of the social challenges technology will bring.

Savulescu ultimately sees human beings as an inadequate species teetering towards oblivion. By focusing on biological solutions instead of social ones (and failing to fully analyse the interconnectedness of the two) he offers an unconvincing surety that individuals will go unharmed as he works to justify the process that will redeem us: technological human enhancement.

Fuller and Lipinska: The Proactionary Imperative

Fuller and Lipinska (2014) give humans a more elevated standing in their advocacy for transhumanist interventions, recognising them as 'aspiring deities' (2014: 9). They acknowledge their own non-conformist Christian faiths play a vital role in underpinning their transhumanist fervour and warn that a rational analysis of history would render faith necessary to justify any transhumanist endeavours. They argue that if you believe in the cogency of human enhancement, 'yet you see yourself as somehow poised "against" or "beyond" theology, then you need to question the source of your confidence in humanity's indefinite self promotion' (2014: 8). In other words, a belief in God is necessary to justify the faith that transhumanist aims will not end in catastrophe.

At the heart of Fuller's belief system (abetted by Lipinska in *The Proactionary Imperative*) is that we are obliged to pursue technoscientific progress relentlessly and endlessly to reach our god-like destiny or infinite power, effectively to serve God by becoming God. This conception of humanity's metaphysical condition lies at the heart of Fuller and Lipinska's justification for what they call the 'proactionary imperative', an ideological extension of the 'proactionary principle' which they identify as 'the 18th century Enlightenment idea of progress on overdrive' (2014: 9). The implications of such a mantra are stark:

> … replacing the natural with the artificial is so key to proactionary strategy…some proactionaries speak nowadays of 'black sky thinking' that would have us concede…the long-term environmental degradation of the Earth and begin to focus our attention on space colonisation. (2014: 99–100)

Fuller and Lipinska fail to realise that even if we are to take their long-term, transcendent aims seriously, those aims are completely undermined by a laissez-faire attitude to social, economic and environmental risk. A sustainable environment is a necessary platform on which to build transcendent goals. Only blind faith and ideological dogma can cast these two objectives (a sustainable environment and transcendent goals) into a logical contradiction. Furthermore, it is not just the Earth itself that would be allowed to fall victim to Fuller and Lipinska's transhumanist super-experiment, but any person or thing that may represent grist for the mill in the transcendent journey.

The full extent of suffering Fuller and Lipinska would be willing to gamble in their cosmic casino is revealed when we analyse what their project would mean for individual human beings: 'a proactionary world would not merely tolerate risk-taking but outright encourage it, as people are provided with legal incentives to speculate with their bio economic assets. Living riskily would amount to an entrepreneurship of the self' (2014: 132). The horrors that are likely to be endured by this globalised market of human experimentation are conceived of as mere learning experiences as 'proactionaries… seek large long-term benefits for survivors of a revolutionary regime that would permit many harms along the way' (2014: 101). Progress on overdrive will thus require sacrifices. The economic fragility that humans may soon be faced with could prove extremely useful to the proactionary goals. In a society where vast swathes of the population are entirely reliant on the State for survival, market forces would determine that the less the State provides, the more people are likely to risk for a lower reward. Hence, 'proactionaries would re-invent the welfare state as a vehicle for fostering securitised risk taking' (2014: 42) and 'the proactionary state would operate like a venture capitalist writ large' (2014: 42). Social justice and the promotion of equality are not a concern of the state here:

> To the proactionary, there is nothing intrinsically valuable in...'equality', despite its reputation as a postmodernist utopia. On the contrary, it looks like the enforcement of... a 'fragile' approach to ecology that fails to recognise the creative power of destruction in both natural and human history. (Fuller and Lipinska 2014: 3–4)

At the heart of this thinking is the removal of basic rights for 'Humanity 1.0', Fuller's (2010) term for modern, non-augmented human beings, and the replacement of it by duties. These, in effect, constitute rights for the future augmented transhuman entity as they force the current manifestation of human beings to make sacrifices with the intent of increasing the likelihood of enabling this 'Humanity 2.0' to come into being.

The duties that constitute future humans' rights include that the very code of our being can and perhaps must be monetised: 'Conceptualise our genetic material as property that one is entitled, and perhaps even obliged, to dispose of as inherited capital' (Fuller and Lipinska 2014: 32) and 'personal autonomy should be seen as a politically licensed franchise whereby individuals understand their bodies as akin to plots of land in what might be called the "genetic commons"' (Fuller and Lipinska 2014: 69–70). The neoliberal preoccupation with privatisation should thus extend to human beings according to *The Proactionary Imperative*. Indeed, the life-time of debt that is the lived reality of most citizens in developed advanced capitalist nations takes a further step as you are born into debt: 'Simply by virtue of being allowed to live, you are invested with Capital on which a return is expected' (2014: 107). It is something of an understatement when they claim, 'Clearly employed here is a radicalisation of attitude toward the human' (2014: 131). This after all amounts to the privatisation of the human being. But this is not logically too great a leap from current systemic logics: 'However specified, the ultimate goal in this bio-capital utopia is maximum productivity—making the most of one's inheritance' (Fuller and Lipinska 2014: 70–71). Thus the logic echoes the justification within advanced capitalism for its business ontology. Socially moribund masses forced to serve the technoscientific super-project of Humanity 2.0 utilises the ideology of market fundamentalism in its quest for perpetual progress and maximum productivity. The only significant difference is that the stated aim of 'god-like' capabilities is overt, as opposed to the undefined end determined by an ever more efficient market logic, which is needed simply to serve capitalism's endless growth fetish. They further advocate bringing about a war economy model impressed as they are by 'unprecedented feats of innovation by virtue of having to respond to a comprehensive yet unpredictable external foe' (2014: 106). Hence they wish for a kind of permanent shock doctrine, disaster capitalism on acid, whereby 'the prospect of ecological collapse, epidemics or even global financial meltdown [might] serve a similar function to focus minds in our own day' (2014: 106).

Fuller and Lipinska unabashedly embrace the eugenic nature of their ideology: '"transhumanism"—at least in name—owes its very existence to eugenics, whose spirit it continues to promote under the slightly more politically correct rubric of "human enhancement"' (2014: 64). And indeed, in a distant echo of Savulescu's rejection of privacy in favour of security, Fuller and Lipinska acknowledge that 'eugenics requires mass surveillance and experimentation, with the understanding that many in retrospect may turn out to have been used or sacrificed for science' (2014: 63). However, like a PR brochure for corporate responsibility in the age of mass human experimentation, they assure us:

> Yes, this is eugenics, but [not] the classical state authoritarian version… Rather, hedgenetics would be a kind of participatory eugenics, a democratically accountable, legally binding version of eugenics written into the heart of intellectual property law and the regulation of financial transactions. (2014: 128)

How Fuller and Lipinska ensure democratic accountability across the planet or what takes precedence when democracy threatens to limit productivity they do not say. Likewise, there is no mention of the implications of the significant social and economic inequities that would arise from such a state of precarity in terms of the democratic deficit caused by such skewed power relations. Presumably, genocide would only be a serious moral concern for proactionaries if it inhibits progress towards the desired transcendent aim—perhaps by reducing the number of people available for experimentation. It is worth reflecting once more on Savulescu's simplistic notion of 'moral upgrade', and the contestability of the term morality. Such a 'moral upgrade' for Fuller and Lipinska would simply mean making humans more compliant and amenable to whatever facilitates the greatest efficiency in realising our transcendent trajectory.

Whilst Fuller and Lipinska's controversial tract may not be something most transhumanists would advocate in full, it is important nevertheless, as it does represent a rational transhumanist ideology, building on existing market logics, and legal and political structures. Its grain runs more aligned to advanced capitalist logic than against it, only adding an external overriding moral impulse (infinite progress towards omnipotence), which directs all vested interests within the system. Whilst Fuller and Lipinska would argue that they dignify humanity with a superior, indeed divine purpose in the Universe, the implications for 'Humanity 1.0' (unaugmented modern-day humans) would be devastating and deeply dehumanising.

The reason for analysing the transhumanist ideas of Savulescu and Fuller is not just to show how individuals would be under significant personal threat in the social worlds proposed by both thinkers but to establish how each, in very different ways conceptually 'dehumanises'. For Savulescu, humans are evolutionary relics—morally stunted entities ill-fitting for the modern world. Whilst it is hard, in many ways not to agree with him given the parlous state we find ourselves in, it should be noted that Savulescu puts the blame squarely within the moral core of each and every individual, all of whom need fixing. He offers little recognition of the social contingency of human moral failings. We are biologically, not systemically damned. We all offer such a threat to ourselves and each other that we cannot possibly be trusted and must give up all rights to privacy: the fault is with us, not the technology that heightens our potential for destruction or the social systems that enable it, whilst shaping our morality. For Fuller, we are nascent gods—but this conception too renders us 'dehumanised' as our current manifestation is simply a disposable stepping stone on the path to divinity. Humanity is an iteration—an early one at that. We are Humanity 1.0, ripe for brutal and persistent experimentation and overdue an upgrade.

Conceptions of the Human in Neuroscience and Cybernetics

Cultural conceptions of humanity are important. Concepts can inform material reality just as capitalism's ability to turn nefarious ideas and inert matter into priced and categorised entities has become globally manifest. Dominant conceptions of what it means to be human will therefore come to play out in the material reality of what we are to become, though refracted and filtered through the other systemic influences that interconnect in shaping our development. Whilst both neuroscience and cybernetics are not monolithic entities, they contain certain normalising cultural practices. Both are involved in the development of practical technological advances and thus the implicit assumptions and conceptions they carry are likely to have real-world impacts. Whilst neuroscience and cybernetics have competing starting points for their conception of the human (the former fundamentally material and the latter essentially bodiless), their interests begin to converge. Both open up areas of possibility for transhumanist aims and technologies, and the logics of each have potentially 'dehumanising' implications inherent to their reductionist aims and interpretations.

The conscious human mind is perhaps the most complex object in the known universe and its greatest mystery. The quest to explain it is most readily achieved through a reductive conceptualisation of it as a simple, explicable, material entity. This process is well underway: 'as Crick provocatively put it, locating the seat of free will in the anterior cingulate sulcus… "You are nothing but a bunch of neurons"' (Rose and Rose 2016: 18). There is a revealing dual and contradictory pull within contemporary neuroscience. On the one hand, it is embedded within the logics of advanced capitalism, and it is thus often dependent upon a neoliberal ideological underpinning to ensure its funding and the practical application of its findings in shaping social policy. The result is a focus on individual minds, decontextualised from the social worlds they inhabit as 'Neuroscience's methodological focus on the individual brain is in accord with that of neoliberalism on the individual rather than the collective' (Rose and Rose 2016: 152). Hence neuroscience offers up the cold neutrality of science to potentially both 'diagnose' and 'fix' imbalances in those who do not function effectively in capitalism's hyper-competitive and consumptive social setting. This reductionism also chimes with the ideal neoliberal subject believing in their own perfectibility through consumerism; an instrumental approach to the self. The social context is discounted as the cause of mental problems if we can see and alter the material manifestation of the problems themselves as neuroscience ultimately promises to do. 'Each "neuro-self" [is] responsible for their own well-being, sustained through the promises of personalised medical care' (Rose and Rose 2016: 10). Thus, there is no need to 'mitigate the inequality and deprivation that are an integral part of an intensely marketised economy' (Rose and Rose 2016: 155).

The contradictory pull takes us beyond the analysis of the decontextualised, non-social, individual. In defeating the mystery of the mind through its ultimate materialisation, the individual is no longer an individual at all, but rather is reduced 'to collections of neurons (nerve cells) and synapses (the junctions between them)' (Rose and Rose 2016: 10):

> The task of [neuroscience] is thus to elucidate the genetics, biochemistry and physiology of the brain processes and, in doing so, to make the mind, and the person it inhabits, merely a 'user illusion', fooling people into thinking that they are making decisions whereas it is really the brain that is doing it. (Rose and Rose 2016: 19)

This schism between the fetishised, atomised individual that is integral to the modern ideological incarnation of capitalism and the deconstruction of the notion of the 'liberal' individual that science tends towards with its materialisation of the mind (and indeed all matter) thus indicates a fundamental disconnect.

As neuroscience seeks to locate the material incarnation of such immaterial concepts as 'experience' and 'memory', cybernetics takes these immaterial concepts and denies them any materiality at all. In both directions, an aspect of the human disappears—for neuroscience our conscious experiences, our identities are chimeras—nothing more in reality than a physical interaction of tiny material entities. For cybernetics, we are only information, but whatever the material on which this information exists is an irrelevance and the essential information can be transferred to or carried by other materials or exist independently of any material at all. We are thus minds without bodies or bodies without minds: but not a mind inescapably embedded in a body. Henry Markram (in Rose and Rose 2016), initiator and coordinator of the EU's $1.2 billion *Human Brain Project* states, that 'the human brain is the world's most sophisticated information processing machine'. This statement reveals a further implication of 'ruthless reductionism'. Not only does the individual disintegrate under the pressure of definition by constituent parts, but the purpose of the individual becomes reconstituted too. Humans become 'information processors' and 'machines'. Indeed transhumanist thinking is replete with language reconstituting humanity in mechanic terms, especially the language of computers.

In fact, this metaphor of the mind as an information processor has a lineage that stretches well beyond Henry Markram, as Katherine Hayles painstakingly shows in *How We Became Posthuman*. Hayles states,

> At the inaugural moment of the computer age, the erasure of embodiment is performed so that "intelligence" becomes a property of the formal manipulation of symbols rather than an action in the human life-world… Aiding this process was a definition of information that conceptualised information… as an entity distinct from the substrates carrying it. From this formulation, it was a small step to think of information as a kind of bodiless fluid that could flow between different substrates without loss of meaning or form. (1999: xi)

Drawing upon a loose history of cybernetics, Hayles seeks to show that 'such abstractions as bodiless information' (1999: 12) are contingent upon historical assumptions. In analysing this history and revealing its contingency, Hayles attempts to undermine the process that has come to create an 'information/materiality hierarchy' (1999: 12) in which information takes precedence and can exist essentially independently of any material instantiation.

We have a long history of conceiving of ourselves and everything else in metaphors which reflect our contingent and transitory worldview. But as Grasse points out, this is highly problematic:

> The brain, much like the genome, does absolutely nothing by itself. When separated from this messy matrix of embodied relationships... the brain really has no capacities at all, computational or otherwise. In a profound ontological sense, the human body-brain-mind-spirit is an emergent phenomenon in which the whole is "exponentially" more than the sum of its parts... Reductionistic approaches to the neurosciences and artificial intelligence can quickly become category mistakes, as silly as trying to use particle physics to interpret a novel, evolutionary psychology to ascertain the truth-value of physics, or life-extension technology to realise the meaning of life. Metaphors matter. (In Tirosh-Samuelson and Mossman, (eds.) 2005: 475–6)

These category mistakes and cultural narratives play a role in forming our material reality as Hayles recognises:

> When I say virtuality is a cultural perception, I do not mean that it is merely a psychological phenomenon. It is instantiated in an array of powerful technologies. The perception of virtuality facilitates the development of virtual technologies, and the technologies reinforce the perception. The feedback loops that run between technologies and perceptions, artefacts and ideas, have important implications for how historical change occurs. (1999: 14)

As such, the paradigmatic interpretation of the mind as a bodiless information processor may bring about the creation of technologies that reflect this ideological standpoint. In Lanier's efforts to humanise the information economy he argued: 'Digital information is really just people in disguise' (2013: 15). However, Hayles shows us that in cybernetic thinking the sentence can easily be flipped: people are really just digital information in disguise. Hayles' concern is not just the abstract, conceptual, dehumanising notion that humans are essentially bodiless algorithms, but the implications of this assumption made manifest in real-world technologies.

Just as the conceptual assumption of the hierarchical supremacy of information over embodiment may have material implications for technological developments, so too will more direct economic interests. The *Human Brain Project* is a case in point:

> The innovative technologies BRAIN funds, often seemingly remote from studies of the brain but of wide industrial potential, range from nanoparticles to optoelectronics. Unlike the HBP, the US project acronym makes clear its technological and wealth creating intentions. Significantly, BRAIN is being funded not only by the US Federal Agency and the National Institutes of Health, but also by the military through DARPA. (Rose and Rose 2016: 48–9)

The 'innovative technologies' that come out of such a project cannot help but have ends tied to the interests of those organisations that fund it: notably military and financial. It is in particular the combination of these culturally influential factors: military and financial interests, coupled with an undermining of the recognition of the human as a specific and embodied entity, that could lead to a particularly concerning trajectory. The death of the liberal human subject, though in itself something that may be welcomed, does not necessitate its replacement with a more benign cultural conception. Ideas of the human as an evolutionary stepping stone, as inherently biologically destructive, as an accident of meaningless material interactions or as nothing but an information processor, leave human beings potentially subject to abuse, especially when it is threatened by a feast of technological innovations themselves financed by interests focused on profit or war.

Religion and Colonialism: Transcendent Promises of Bodiless Information

The equating of the human mind to mere information has a further significance. It offers transhumanists such as Moravec (1990, 2000) and Kurzweil (2000, 2005, 2010) an escape from the myriad limitations of embodied existence. Hayles explains, 'Information viewed as pattern and not tied to a particular instantiation is information free to travel across time and space... it can be free from material constraints that govern the mortal world... we can achieve effective immortality' (1999: 13). The idea that information itself is the fundamental essence, and its materiality is merely a temporary, contingent and unnecessary factor is a recurring assumption in many of the descriptions of the future Kurzweil and Moravec conjure. Kurzweil's imaginary 2099 is composed of machine based-humans, made not of 'wetware'—flesh and blood—but rather software: an electronic or photonic basis for cognition. No physical body is required, but can be conjured if desired either through virtual reality or through the creation of 'reconfigurable nanobot swarms' (e.g. in Garreau 2005). There are various aspects of this conception that are appealing—our lack of dependence on a physical body is suggestive of immortality, such fluid physicality as 'reconfigurable nanobot swarms' is a liberating idea compared with the inhibiting nature of our current mortal embodiment. Infinite plasticity of form (including absence of any material form) also renders Nature endlessly abundant. Whatever can be imagined can be made manifest. As Thorpe explains: 'Kurzweil's fantasy...denies limits by imaginatively escaping from the constraints of Earth as the relevant environment...the rejection of spacial limits (of Earth) and temporal limits (via radical life extension) is achieved through decorporealisation' (2016: 79). However, to believe in this requires certain metaphysical leaps of faith. Hayles, brings us back to Earth, quite literally by rightly asserting,

> In the face of such a powerful dream, it can be a shock to remember that for information to exist, it must always be instantiated in a medium... The point is not only that abstracting information from a material base is an imaginary act but also, and more fundamentally, that conceiving of information as a thing separate from the medium instantiating it is a prior imaginary act that constructs a holistic phenomenon as an information/matter duality. (1999: 13)

Kurzweil's dream is not science; it is religion. As if to emphasise the point, Meghan O'Gieblyn enlighteningly displays the similarity between Kurzweil's conception of history and certain interpretations of the Bible:

> Like the theologians at my Bible school, Kurzweil… had his own historical narrative. He divided all of evolution into successive epochs. We were living in the fifth epoch, when human intelligence begins to merge with technology. Soon we would reach the "Singularity", the point at which we would be transformed into what Kurzweil called "Spiritual Machines". We would transfer or "resurrect" our minds onto supercomputers, allowing us to live forever. (2017)

Here, religion is playing a different role than it does for Fuller and Lipinska. Whereas for them it is the justification for an irrational faith in progress, for Kurzweil it is almost a rhetorical device. Kurzweil is promising the spiritual (and material) benefits of religious salvation. Indeed, these most transcendent versions of transhumanism, where consciousness, or perhaps just intelligence, leaves the body and exists in some virtual manifestation, require an updated version of the 'soul'. Thus for Kurzweil a person can be conceived of as a pattern of energy. This is certainly a divergence from the 'ruthless reductionism' of neuroscience:

> A pattern, transhumanists would insist, is not the same as a soul. But it's not difficult to see how it satisfies the same longing. At the very least, a pattern suggests that there is some essential core of our being that will survive and perhaps transcend the inevitable degradation of flesh. (O'Gieblyn 2017)

The great irony about this technological salvation is that 'What makes the transhumanist movement so seductive is that it promises to restore, through science, the transcendent hopes that science itself has obliterated' (O'Gieblyn 2017). It cannot make these promises without slipping back into the religious myths from which science seeks to liberate us. Even technology itself often becomes imbued with a kind of animism that renders it 'not only autonomous but also, in a sense alive, and perhaps a supernatural power' (Thorpe 2016: 98). That a human individual can never become a 'nanobot swarm', and in some intelligible sense remain the same individual, does not negate the possibility of humans being superseded by nanobot swarms. Gray recognises that in Kurzweil's pseudo-religious promise of eternal life, there lies only annihilation as far as the human individual is concerned: 'the individual mind is uploaded into a virtual universe. A speck of humanity becomes part of a cloud of consciousness or information.

Whatever survives, the individual is extinguished. Death is not conquered but triumphs unnoticed' (Gray 2011: 218).

Despite this, Kurzweil confidently declares that 'intelligence' will conquer the universe:

> The law of accelerating returns will continue until nanobiological intelligence comes close to 'saturating' the matter and energy in our vicinity of the Universe with our human-machine intelligence... Ultimately, the entire universe will become saturated with our intelligence. This is the destiny of the universe. (Kurzweil 2005: 29)

This is very much colonialism, potentially quite literally, on drugs. Transhumanist pursuits are not just a threat to vast swathes of human individuals. It is an existential threat to the species as a whole because the information with which Kurzweil wants to colonise the universe cannot be human intelligence in any meaningful sense. As Gray understands,

> Immortalism is a program for human extinction, a vanishing act more complete than any that seems likely in the natural course of events...In the immortalist scenario humans engineer their own extinction: intervening in the evolutionary process to create a new species, the animal that yearned to live for ever puts an end to its own existence. (2011: 218–9)

Gray's Stoicism perhaps underplays the devastation that Kurzweil aims to wreak. It is not just human extinction that these pretensions of universal colonisation augur, but all organic life. These intents are colonialist in nature and of the most extreme kind. They are also supported by religious myths promising eternal salvation: a disturbingly familiar story used to justify expansionist aims and domination.

Christopher Coenen argues this colonialist strain has a distinct lineage in transhumanist thought. He claims Winwood Reade's book *The Martyrdom of Man* (1872) 'developed the blueprint for the ideological nucleus of modern transhumanism by creating a specific set of visions of and a narrative about the future of humankind' (Coenen 2014: 41). These include space colonisation, the promise of a new human corporality, the idea of humanity functioning as a hive mind, immortality, and the conviction that humanity will come to rule the universe as a God-like posthuman entity (Reade 2004). Reade's vision is also based on a teleological notion of progress, buoyed by the ever-expanding breadth of human understanding due to the accumulation of rationally construed knowledge. However, it is not just this enlightenment inheritance that is evident in Reade's thinking. Coenen points out that his ideas 'reflected the imperialist context of his life and activities…providing a grand narrative in which all past human endeavours and British imperialism in particular were presented as steps towards a grandiose future' (2014: 41). He concludes 'the genesis of transhumanism has been influenced by the notion of an "empire" and shaped by the imperialist reality of the late nineteenth and early twentieth centuries' (Coenen 2014: 41). Furthermore, Coenen traces a history of transhumanism as 'an expression of displaced eschatological needs' (2014). Thus transhumanist ideas were part of a wider cultural shift which drew upon the awesome scope of nature as inspiration for a sense of the sublime. Coenen states:

> During the nineteenth century, gradualist geology, Darwinianism and cosmology expanded the time horizons of modernity in both directions. The distant past and the far future became subjects of inquiry and speculation. The awe-inspiring timescales and vastness of the universe created a new urgency of the mathematical sublime. (2014: 39)

This was part of an attempt to dignify humanity in a way that makes us part of and perhaps integral to the 'new insights into the immenseness of timescales and vastness of space' (Coenen 2014: 39). Technoscience is sanctified as the methodology for the realisation of a new human self-assertion required after Darwin's strike against human narcissism. Replacing eschatological fantasies with colonialist pretensions is deeply embedded in the cultural history of transhumanist thought.

Genocide: The Incipient Event

A significant proportion of genocide scholarship constitutes an endeavour to define a 'universal essence' among the diverse instantiations of the crime. It attempts to root through commonalities and distinctive traits to reach an overarching conception of its particular nature and to determine the necessary conditions required to generate such a state. This train of thought often leads to over-simplified and stereotyped versions of what genocide looks like. Particular focus is given to the role of despots in control of states. However, there is an increasing awareness of the complexity of genocide as a concept, and resultantly, broader and more nuanced reconceptualisations of its defining features. Louise Wise considers genocide from an 'ecological' perspective:

> as a contingent social (and material) construct that varies in form, dynamics, and manifestation across different historical and geographical contexts, instead of looking to uncover its 'universal form. '...we should accommodate an understanding of genocide as a set of interacting processes, actors and relationships that evolves, or 'becomes,' over time, eschewing explanatory reliance on linear and simplistic, personality-driven notions of 'mad or bad' leaders...we need to move beyond the analytical confines of the state by facilitating perception of how episodes of genocide are situated within broader global contexts. (2015: 256)

She argues that 'genocide does not "erupt," rather, it becomes over time, or at least the conditions of its emergence, or its potentiality, becomes over time' (2015: 257). Ultimately through this ecological framework, she defines it as a 'complex system that produces widespread social death' (2015: 256). When genocide is understood this way it becomes clear, that many of the emergent forms of dehumanisation discussed here could come to represent genocidal potentialities.

The gulf between rich and poor could be vastly widened by radically powerful technologies. When we consider the introduction of highly effective psycho-pharmaceuticals, genetic modification, nanotechnology, robotic prosthetics, brain to computer interfaces, superintelligent artificial intelligence and the development of life expansion, the inequities we are already seeing will surely be magnified if such a world is born out of the competitive nature of capitalism with its inherent collisions of class interests. The already discredited idea of trickle-down economics can not even begin to apply when species divergence exists. Peter Frase's Four Futures imagines a scenario he labels 'Exterminism' in which hierarchy and scarcity co-exist in a world of powerful technologies:

> A world where the ruling class no longer depends on the exploitation of working-class labour is a world where the poor are merely a danger and an inconvenience. Policing and repressing them ultimately seem more trouble than can be justified. This is where the thrust 'towards the extermination of multitudes' originates. Its ultimate endpoint is literally the extermination of the poor so that the rabble can finally be brushed aside once and for all leaving the rich to live in peace and quiet in their *Elysium*. (2016: 126)

What Michael Sandel (2009) refers to as the 'Skyboxification' of life, namely the increasingly segregated social worlds of the elites in developed countries, is already with us. The falling share of the labour to capital ratio continues apace and shows no sign of abating as automation unemployment becomes increasingly evident. The logics of advanced capitalism decry 'social security' and seek to privatise as much of the economy as possible. The treatment of ever-larger numbers of surplus populations, most notably refugees, has become increasingly dehumanising. Public discourse that encourages antagonism towards such groups has fanned the flames of a recent wave of 'populist' politics in Europe and America. Meanwhile, the increasingly powerful technologies that enable this parlous and precarious state of affairs, and render their systemic logics ever more opaque, continue to develop at an exponential rate.

For all its failings and its anthropocentric concessions, liberal thinking underpins much of the protections that are culturally and legally embedded within our political and legal frameworks. Human rights are not the natural state of affairs given human history, but a hard fought and profoundly important achievement. The fragmenting of the very concept of 'humanity' undermines the humanist traditions in which these achievements have developed. As Francesca Ferrando explains, transhumanists and posthumanists 'share a common perception of the human as a non-fixed and mutable condition, but they generally do not share the same roots and perspectives' (2013: 2). For transhumanists, the mutability represents the possibility for enhancement—opening up a teleological narrative of evolution towards an upgraded transhuman entity. For posthumanists, it represents a fracturing of the liberal human subject and an undermining of its hegemonic principles. As Elaine Graham argues,

> This generates an imperative to interrogate more deeply the values and interests that underpin any representation of the 'posthuman condition'. What is at stake, supremely, in the debate about the implications of digital, genetic, cybernetic and biomedical technologies is precisely what (and who) will define authoritative notions of normative, exemplary, desirable humanity into the twenty-first century. (2002: 11)

Echoes of this sentiment can also be found in Andy Miah's (2008) analysis. For him, posthumanism is 'the study of the collapse of ontological boundaries...of how moral landscapes might be transformed by this occurrence' (2008: 96) and 'posthumanism is a philosophical stance about what might be termed a perpetual becoming' (2008: 98). It is this perpetual becoming that requires 'the ongoing re-definition of an "ethics of bodies that matter"' (Miah 2008: 94). The 'human' then in posthumanism and transhumanism lacks essentialism and stability—the 'post' and 'trans' prefixes clearly point to this. But whereas the 'post' demands a deep ethical consideration of the implications of the unmooring, the 'trans' tends to claim a clear and advantageous direction of travel. The process of reconceptualising the human in materially reductionist terms, bodiless virtual terms and even as commodified capital, would suggest the posthumanist conception is a considerably more appropriate position. Shaw articulates the stakes effectively:

> As life itself, understood as bio-genetic information, becomes commodified, it equally becomes manipulable and hackable. Understandings of what constitutes corporeality, consciousness and individuality have always been contested, but now they emerge as significant stakes in projected re-definitions of legal personhood and considerations of rights accorded to species other than human. (2015: 1)

This highlights the necessity for ethical consideration of the mutable, untethered human condition, rather than simplistic, deterministic and teleological narratives. It also indicates that existent notions of human rights may be undermined by challenges to our conception of the human.

A further concern emerges from Wise's recognition of 'a "homology" between colonialism and genocide' (2015: 260). The hyper-colonial pretensions of Kurzweil's aims for human-descendent intelligence and Fuller's for god-like universal dominance are unnerving given the interconnectedness Wise perceives. Like many colonial pursuits, religion plays an overt role in Fuller's ideology, his faith a justification for his willingness to run roughshod over any legislation deemed precautionary—that is to say with a focus on stability and sustainability. The religious elements of Kurzweil's vision are less overt, but its roots in Christian eschatology and its reliance on metaphysical positions beyond the realm of science render it dependent on faith. The determinism of their views also points towards faith, not rationality. For Moravec, modern humans do not matter because they are

> going to be left behind, like the second stage of a rocket. Unhappy lives, horrible deaths, and failed projects have been part of the history of life on Earth ever since there was life; what really matters in the long run is what's left over. Does it really matter to you today that the tyrannosaur line of that species failed? (Cited in Thorpe 2016: 109)

He explicitly links this 'progress' to prior colonial pursuits (1990). To Moravec colonialism, genocide, ecological devastation are all just part of the natural history of progress: civilising forces in the grand march towards our intelligent designs colonising the universe. Cudworth and Hobden insightfully ask:

> Where… do these Western framings leave the concept of civilisation?… the development of such a discourse drew on a particular conception of what it meant to be human, in which being properly human demanded a separation from the rest of nature, constructed on the capability to control and exploit. The civilising mission of European colonialism was to both exert cultural superiority and subject 'barbaric' cultures to a form of uplift. (2017: 123–4)

Moravec's worldview mirrors the 'civilising mission of European colonialism' to which Cudworth and Hobden refer. Again the form of 'uplift' is present (in the guise of human enhancement); again the barbarism is justified through a supposed cultural superiority (in the form of advanced technology).

A further indicator of genocidal potential is access to genocidal tools. As Savulescu raised earlier (as a justification for doing away with liberal democracy and replacing it with a panoptical authoritarianism), it is impossible to conceive of developing radically powerful technologies without developing radically powerful weapons. Gray characterises it thus:

> At the start of the 21st century technologies of mass killing have become more powerful and more widely dispersed. Not only nuclear weapons but also chemical and biological weapons are steadily becoming cheaper and more easily usable, while genetic engineering is sure to be used to develop methods of genocide that destroy human life selectively on a large-scale. In a time when the spread of knowledge makes these technologies ever more accessible death rates could be very high, even among those whose longevity has been artificially enhanced. (2011: 210)

Taken on its own, the sheer potential accessibility of weapons of such magnitude may represent a catastrophic danger. Coupled with the logics discussed above, the threat becomes all the more significant. A world of highly accessible weapons of mass destruction may also provide elite groups with a potential justification to take extreme measures to defend themselves.

The systemic dehumanising logics of advanced capitalism when applied to the incipient and emerging dynamics of radical technological possibilities may not as yet be manifesting anything like 'patterns of killings' that would invoke genocidal fears. However, the point at which such an event may occur could, unlike all prior cases of genocide, result in an unstoppable and total process. The cocktail of deadly power, immeasurable opportunity and inhuman pace of change mean that the conventional 'early warning' signs that Wise mentions would come far too late. It is vital then to recognise, despite the evident complexity of the process of realising transhumanist potentialities, the inherently genocidal latent tendencies that underpin much of the philosophy in the context in which it is emerging. The existential dangers of runaway computer intelligence and related fears are widely covered, but the inchoate genocidal propensities, much less so.

References

BMJ, (2017) "Effects of health and social care spending constraints on mortality in England: a time trend analysis", available at: https://bmjopen. bmj. com/content/7/11/e017722 (Accessed 24th April 2019).

Brynjolfsson, E. & McAffe, A. (2014) *The Second Machine Age*, New York: WW Norton & Co.

Carr, N., (2015) *The Glass Cage*, London: The Bodley Head

Chace, Calum (2016) *The Economic Singularity*, London: 3Cs.

Coenen, C., (2014) "Transhumanism and its Genesis: The Shaping of Human Enhancement Discourse by Visions of the Future" in *Humana Mente Journal*, Vol. 26, p. 35–58

Cudworth, E. & Hobden, S., (2017) *The Emancipatory Project of Posthumanism*, Oxon: Routledge

Dorling, D., (2014), *Inequality and the 1%,* London: Verso

Ferrando, F., (2013) "Posthumanism, Transhumanism, Antihumanism, Metahumanism, and New Materialisms" in *Existenz*, Volume 8, No 2, Fall 2013, available at: https://existenz. us/volumes/Vol. 8–2Ferrando. pdf (Accessed 24th April 2019)

Ford, M., (2015) *The Rise of the Robots – Technology and the Threat of Mass Unemployment*, London: Oneworld Publications.

Frase, P., (2016) *Four Futures: Life After Capitalism*, London: Verso.

Fuller, S. (2010) *Humanity 2.0 What It Means to be Human, Past, Present and Future*, Basingstoke: Palgrave Macmillan

Fuller, S. & Lipinska, V. (2014) *The Proactionary Imperative, A Foundation for Transhumansim*, Basingstoke: Palgrave Macmillan.

Garreau, J. (2005) *Radical Evolution*, New York: Broadway Books.

Graham, E. (2002) *Representations of the post/human*, Manchester: Manchester University Press.

Gray, J. (2011) *The Immortality Commission*, London: Granta Books

Harari, Y. N., (2016) *Homo Deus*, London: Harvill Secker.

Harari, Y. N., (2017) "The Rise of the Useless Class", available at: http://ideas. ted. com/the-rise-of-the-useless-class/ (Accessed 24th April 2019).

Hayles, K. , (1999) *How We Became Posthuman*, London: University of Chicago Press.

International Organization for Migration, (2017) "Global Migration Data Analysis Centre: Data Briefing Series", Issue No. 8, March 2017, Available at:

http://publications. iom. int/books/global-migration-data-analysis-centre--data-briefing-series-issue-no-8-march-2017 (Accessed 7th June, 2017).
Jacobs, M. , & Mazzucato, M., (2016) *Rethinking Capitalism*, Chichester: Wiley Blackwell.
Kurzweil, R., (2000) *The Age of Spiritual Machines: When Computers Exceed Human Intelligence*, London: Penguin.
Kurzweil, R. (2005) *The Singularity Is Near*, New York: Penguin.
Kurzweil, R., (2010) *Transcend*, Pennsylvania: Rodale Press.
Lanier, J. (2013) *Who Owns the Future?*, London: Allen Lane.
Miah, A., (2008) "A Critical History of Posthumanism", in: Chadwick, R. & Gordijn, B. Medical Enhancements and Posthumanity. Springer, pp. 71–94.
Moravec, H., (1990) *Mind Children: The Future of Robots and Intelligent Machines*, Massachusetts: Harvard University Press.
Moravec, H., (2000) *Robot: Mere Machine to Transcendent Mind*, Oxford: OUP.
More, M., Vita-More, N. (eds.) (2013) *The Transhumanist Reader*, Chichester: Wiley-Blackwell.
O'Gieblyn, M., (2017) "God In The Machine: My Strange Journey Into Transhumanism" in *The Guardian*, available at: https://www.theguardian.com/technology/2017/apr/18/god-in-the-machine-my-strange-journey-into-transhumanism (Accessed 27th June, 2017).
Oxfam, (2018) *Reward Work Not Wealth*, available at: https://www.oxfam. org/en/research/reward-work-not-wealth (Accessed 1st June, 2018).
Oxfam (2019) 5 Shocking facts about global poverty, available at: https://www.oxfam. org/en/even-it/5-shocking-facts-about-extreme-global-inequality-and-how-even-it-davos (Accessed 24th April, 2019).
Persson, I., & Savulescu, J. (2008), "The Perils of Cognitive Enhancement and the Urgent Imperative to Enhance the Moral Character of Humanity", available at: http://www.bep. ox. ac. uk/__data/assets/pdf_file/0003/9390/Finalsubmittedcorrected09april_08. pdf (Accessed 21st May 2017).
Reade, W., (2004) *Martyrdom of Man*, Hawaii: University Press of the Pacific.
Rose, H. , & Rose, S., (2016) *Can Neuroscience Change Our Minds?*, Cambridge: Polity.
Sandel, M., (2009) *What Money Can't Buy: The Moral Limits of Markets*, London: Penguin.
Sassen, S. (2014) *Expulsions: Brutality and Complexity in the Global Economy*, Cambridge: Harvard University Press.
Savulescu, J. (2009) "Genetically enhance humanity or face extinction", Available at: https://www.youtube. com/watch?v=PkW3rEQ0ab8 (Accessed 21st March, 2017).

Savulescu, J. (2012) "Enhancing Equality" in Lipper-Radmussen, K. , Thomsen, M. R. Wamberg, J. (eds.) (2012) *The Posthuman Condition*, Aarhus: Aarhus University Press.
Shaw, D. B. (2008) *Technoloculture*, King's Lynn: Berg.
Shaw, D. B. (2015) *Post-Human? And Who Are We Anyway?*, University of East London.
Srnicek, N. and Williams, A., (2015) *Inventing the Future: Postcapitalism and a World Without Work*, London: Verso.
Srnicek, N., (2017) *Platform Capitalism*, London: Verso.
Tirosh-Samuelson, H., & Mossman, K. L., (eds.) (2005) *Beyond Humanism: Trans- and Posthumanism*, Frankfurt: Peter Lang.
Thorpe, C., (2016) *Necroculture*, New York: Palgrave Macmillan.
Wise, L., (2015) *Social Death in Sudan: Towards an Ecology of A Genocide*, King's College London, University of London.

Part V

The *Cunning of Reason*

Posthumanism, Transhumanism, Superhumanism and Metahumanism from an Adaptive Standpoint

Paul Jorion

When I write "Posthumanism, transhumanism, superhumanism and metahumanism from an adaptive standpoint", what I mean by "adaptive standpoint" is "as far as are concerned the survival of individual humans and of the human species as a whole". The question I'm raising thus is to what extent are posthumanism, transhumanism, superhumanism and metahumanism views that analyse adequately and recommend useful proposals pertaining to safe, long and enjoyable individual lives, and indefinite survival of the human species in optimal circumstances.

Before going any further, let us characterise briefly each of those varieties of anti- or neo-humanism.

Although the term posthumanism refers in most cases to an updated and revised version of the humanism of olden times having shed away one or more of its basic tenets, such as a *speciest* anthropocentric bias, and replaced it by some expansion of the sphere, the term has been used also

P. Jorion (✉)
Department of Ethics, Université Catholique de Lille, Lille, France
e-mail: paul.jorion@univ-catholille.fr

P. Jorion (ed.), *Humanism and its Discontents*,
https://doi.org/10.1007/978-3-030-67004-7_11

to refer to visions of a world from which humankind would have been entirely wiped out.

Transhumanism has adopted the radical view of ignoring what had caused in the mid-twentieth century qualms with the Enlightenment and its utopia of an ever unfolding human perfectibility, therefore reconnecting at the very location where the disconnect had taken place. Thus, my labelling it as "The Enlightenment recovered". Transhumanism offers a discourse justifying, however daring they might be, the steps required to ensure the survival of the human species, whether that would occur on planet Earth or anywhere else in the universe.

As opposed to posthumanism and transhumanism promoting a humankind progressing in lockstep to a new stage, Nietzsche's superhumanism and del Val and Sorgner's metahumanism are elitist.

Nietzsche had said so unambiguously:

> The problem that I set here is not what shall replace mankind in the order of living creatures (—man is an end—): but what type of man must be bred, must be willed, as being the most valuable, the most worthy of life, the most secure guarantee of the future. (Nietzsche 1895 [1888]: 3)

Elitism, similarly, with del Val and Sorgner if, according to them, "Monsters are promising strategies for performing this development away from humanism" (del Val and Sorgner 2010), as there is no way monstrosity may rise to the level of universality without turning automatically into a new normalcy.

Why envisage these four views in terms of adaptation? Because survival of the human race is in jeopardy at the time we speak as it has trespassed the carrying capacity of the environment as far as it is concerned. In addition, its own capacity at maintaining a liveable environment for itself is seriously compromised, due to pollution, climate change, the accumulation of waste being toxic or not, and the exhaustion of natural resources. Although there is no consensus among scientists on this issue, scenarios envisioning the extinction of humankind within the coming three generations or one hundred years are sufficiently plausible to be taken in dead earnest.

Depending on which mammal species one has in mind, individuals live from a couple of years to about one hundred, the latter being our own case, the rule being that ageing takes its toll ineluctably after a mammal has trod along its reproductive time lapse. A mammal species' lifetime is typically 2.5 million years.

Faced now with the possibility of extinction of our own species, we, human beings, display a wide variety of responses whereof I will first produce a brief catalogue before examining where posthumanism, transhumanism, superhumanism and metahumanism stand in that respect.

First option:

I. the denial of an extinction's risk, which might be

I.1 a passive, privately held personal denial or

I.2 an active denial when a person uses one's power and money to finance denial publicly through an *agnotology* campaign, the latter being defined as "culturally induced ignorance or doubt, particularly the publication of inaccurate or misleading scientific data", or to sponsor smears of bona fide scientists or activists, as was recounted in detail by N. Oreskes and E. M. Conway in *Merchants of Doubt* (2010).

Second option:

II.a an acceptance of doom, here also under two guises, a passive and an active one. In the passive form of acceptance of extinction,

II.a.1 a process of mourning for the fate of the human race is at work with an individual holding the fatalistic view that our destiny was sealed, possibly several centuries ago, when, for instance, at the end of the eighteenth century we started tapping massively fossil fuel reserves as energy sources and burning them into the atmosphere for lack of any alternative medium where to send the fumes.

II.a.2 in the active form of an acceptance of extinction, a person not only hails the demise of humankind but contributes to precipitating it as a proactive way for saving whatever part can be of the current remaining biodiversity. Here active doomsayers intend their activism to be a tribute to life's survival, with an emphasis on its most sophisticated manifestations apart from us undignified blunderers on our way out for the sake of all.

As an alternative, or as a complement to acceptance of our own demise, we can contemplate

II.b the transmission of our own legacy to future generations of self-replicating intelligent machines, best able to survive in a type of environment that we have failed to keep friendly towards ourselves and has therefore turned hostile. Such successors of ours won't care about the presence or not of oxygen in the atmosphere or whether or not the water around is drinkable according to human standards of drinkability.

Third option:

III. restoring our capacity at remaining within the boundaries of the carrying capacity of our environment through the deliberate goal of reducing our imprint on planet Earth. Different methods exist for achieving this aim, covering a wide range of types of behaviour, from

III.a individual reform through voluntary frugality, to

III.b eugenics when filtering of future generations' desired features is promoted and implemented or to

III.c *exterminism*, such as genocide, when a wilful shaping is intended of not only future generations and their reproductive potential but also of current ones, starting now.

Fourth option:

IV. expanding the carrying capacity of the human species' environment through technological innovation. This is by now a traditional way for humans to proceed: every one of the green revolutions, the development and rapid expansion of aquaculture, has allowed expanding our species' carrying capacity. The race has however turned into a losing battle as destruction progresses now faster than expansion.

Fifth option:

V. adapting ourselves to a decaying environment, either through

V.a sheltering ourselves within new subterranean or submersed habitats,

V.b through genetic modifications aiming at making us more resilient, or

V.c through downloading our individual selves as software into silicon-based machines relying hence on light as our sole energy source instead of on difficult to maintain assimilable food.

Sixth option:

VI. colonising new environments. As we have by now colonised all existing niches on Earth, even the more inhospitable, the remaining overture is through colonising other worlds, either within our current reach within our own stellar system, the solar system, or in other stellar systems where exoplanets with physical and chemical features similar to those of our own planet, for lack of strictly identical ones, may provide a welcoming environment or would be able to accommodate *terraforming*.

So, to sum up:

I. Denial of extinction

I.1 Private denial

I.2. "Merchant of Doubt" activist denial

II. Acceptance of extinction

II.a.1. Private mourning of humankind

II.a.2. "Good riddance" activist stance

II.b. Transference of our legacy to our IA heirs

III. Malthusianism

III.a. Individual frugality

III.b. Eugenics

III.c. Exterminism

IV. Progress will prevail

V. Survivalism

V.a. Life in the Underground

V.b. Genetic enhancement

V.c. Dematerialisation of the Self

VI. Space colonisation

The four of them: posthumanism, transhumanism, superhumanism and metahumanism, were conceived by their proponents as a "development away from humanism" as Sorgner underlines.

This was indeed the concern at the basis of the rise of these new upcoming views: to depart from the representation of the human subject which had become central to Western culture, by proposing an alternative "anti-humanism" severed from the central tenets of this "Man" as defined within Genesis, the first book of the Christian dogma: Man in the image of God; all other species, animal or vegetal, being subordinated to Man so as to serve his aims. "And God said, Let us make man in our image, after our likeness: and let them have dominion over the fish of the sea, and over the fowl of the air, and over the cattle, and over all the earth, and over every creeping thing that creepeth upon the earth" (*Genesis*: 1.26).

Man has broken the divine covenant for having tasted the forbidden fruit of the Tree of knowledge, a prohibition akin to that in ancient Greek mythology where Prometheus is cursed and tormented for having provided Man with the premier tool of all technology to come: fire. In the Bible's terms: "And the Lord God commanded the man, saying, Of every tree of the garden thou mayest freely eat. But of the tree of the knowledge of good and evil, thou shalt not eat of it: for in the day that thou eatest thereof thou shalt surely die" (*Genesis*: 2.16-17).

"Let us move away from the superstitious ancient view of mankind and move onto an expanded modern representation, better in tune with our contemporary times", such is the motto of the four ground-breaking new varieties of anti-humanism. But have they got the potential for saving us from eradication? That is what I would like to see here clarified.

Although survival of individual human beings has from the beginning been a central concern of transhumanism, none of the four: posthumanism, transhumanism, superhumanism and metahumanism has stated a clear concern for the survival of the human species as such in a time of threatening extinction.

As for superhumanism and metahumanism in particular, they've been concerned not with the fate of the species as a whole but with the fate of some enlightened members of it: those choosing to rally around their particular stance.

I. Denial of Extinction

It goes without saying that denial of extinction is not adaptive to the extent that it would constitute a positive factor in the fight against the looming threat.

I.1 Private denial puts its subject at the forefront of the early victims to come as it goes hand in hand with unpreparedness for the hardships that will be the auguries of the coming overall demise.

I.2. The "Merchant of Doubt" activist type of denial sets its protagonist in the category of those who are not simply passive before the process of extinction but are also precipitating it.

Does any of the four, posthumanism, transhumanism, superhumanism and metahumanism, hold a special relationship with denial and acceleration of extinction? No, as varieties of anti-humanism they are reflective, they propose a view that is critical of humanism and as such constitute an articulate counter-proposal to it. They do not stand on the side of obscuring but of enlightening, however effective they might be in that respect.

II. Acceptance of Extinction

Acceptance of extinction is not adaptive at least to the extent that is only strictly concerned the fate of our own species. Indeed, the threat of extinction requires as a response a proactive attitude implying planning and gathering together specific resources as the means for preventing such a dire outcome.

II.a.1. Anticipated (what else could it be?) private mourning of humankind amounts to a stoic stance if it is a response attained through profound pondering in the light of a conclusion having been rationally reached that there is no known way for preventing extinction.

II.a.2. The "Good riddance" activist stand that the forcible elimination of the human race is the optimal policy in order to save the other species in their biodiversity, should probably be regarded as a central and standard posthumanist stand, maybe the most commendable of all.

II.b. Transference of our legacy to our IA robot heirs is definitely another variety of the posthumanist stand, where humankind would have vanished but not before having transmitted to self-replicating intelligent machines all that we regard worth of being transmitted to future self-aware creatures, being indifferently carbon- or silicon-based.

This posthumanist position needs however to be clearly distinguished from a transhumanist variation on the same theme akin to it: that of the transference of ourselves into a silicon-based substrate. Even if on the day the practical implementations of both might be difficult to tell apart, the underlying view and aims are clearly separable and have arisen with distinct goals in mind: turning the page and passing the baton to our heirs in the first instance, continuing our own existence in a different guise in the second instance.

III. Malthusianism

As such, the purpose of modifying the behaviour of individuals and the composition of the species as a whole so as to bring it back within the natural boundaries of the carrying capacity of the environment is of course adaptive by logical entailment. The issue here as we're all well aware is not so much that of the legitimacy of the aim as that of the ethical status of the means resorted to.

III.a. Individual frugality would no doubt have been the ideal approach for preventing that we ever had to confront the limitations' issue. The difficulty with frugality as a consistent method at the current time is double: the critical mass of people who need to adopt a particular frugal behaviour and the chronology over which—should a critical mass of people be attainable—a significant change would be observable in the restoration of the carrying capacity of our environment. In those respects, although individual frugal behaviour is adaptive, the near impossibility of attaining critical mass for any particular positive behaviour, and the length of time required to reverse current trends, render the individual frugality approach however desirable, ineffective at the time horizon of three generations or one hundred years within which the threat of extinction needs imperatively to be first challenged, then second, dispelled.

Some attitudes associated with frugality, such as vegetarianism or veganism, fall within the range of a particular brand of posthumanism, that which has extended the realm of dignity to higher order animals or to all animals around us. Within philosophy Arthur Schopenhauer was at the forefront of such a fight, who Nietzsche took first as an educator (*Erzieher*) in his *Untimely Meditations* (1874), before reneging on him in the most counterfeit and provocative manner: "Suffering is made contagious by pity [...] Pity thwarts the whole law of evolution, which is the law of natural selection. It preserves whatever is ripe for destruction; it fights on the side of those disinherited and condemned by life; by maintaining life in so many of the botched of all kinds, it gives life itself a gloomy and dubious aspect. [...] it is a prime agent in the promotion of *décadence*—pity persuades to extinction... [...] Schopenhauer was hostile to life: that is why pity appeared to him as a virtue" (Nietzsche [1888] 1895: 7).

Let us observe in passing that posthumanism holds an "imperialist" view according to which humankind as a whole shares a single concept of dignity, ignoring for all practical purposes that different groups of humans hold widely discrepant views of the extension of the "Us". Dignity may indeed be allocated to the entire realm of living creatures—as posthumanism and metahumanism seem to suggest—but also to dismally small groups such as people of one's own ethnicity, sharing the same religious creed as one, or even restricted to one's own clan or family alone.

III.b. Eugenics as distinctive from exterminism aims at shaping future generations so as to encourage a shift in the composition of the human population towards a particular type regarded as especially desirable, in a perspective undoubtedly regarded as "adaptive" by its proponents even if other parts of the population would question the criteria retained, in particular that part of the population displaying the features determined as those to be eradicated.

Superhumanism and transhumanism undoubtedly betray sympathies towards eugenics, even if most of their spokespeople would deny such inclination, apart from Fuller and Lipinska, promoting "hedgenetics" as a "participatory" and "democratically accountable" type of eugenics (Fuller and Lipinska 2014: 128).

Both strands of superhumanism and transhumanism would need to deal however with the fact and would have to come up with some explanation for why upholders of eugenics invoke their views, although not specifically referring to Fuller and Lipinska, finding there tools for supporting theirs.

It is indeed a logical entailment of the view that a particular type of persons belongs to the ideal type that the preponderance of that type would be sought for and encouraged.

Transhumanism has no qualms with its sympathy towards eugenics even if less controversial words than eugenics are preferably used by its proponents such as "enhancement" or "augmentation" or, as we've just seen, neologisms such as "hedgenetics", genetics and prosthetics being the methods of choice for the enactment of that particular facet of human *perfectibility*.

Needs to be mentioned here however some ambiguity in the case of superhumanism to the extent that the superhuman shows disdain towards reproduction. This was the case for Nietzsche himself as a person and no one entertains any serious doubt that the image of the *Übermensch* was tailored by him from an idealised representation of his own self, as comes about transparently in such texts as *The Antichrist* ([1888] 1895) or *Ecce Homo* ([1888] 1908).

The metahuman "monster" doesn't seem to show much more reproductive proclivity and it could be claimed therefore that both the superhuman and the metahuman pertain to the category of private mourning of the dismal human experience rather than to a Malthusian view consistent with the restoration of harmony between the human race and the carrying capacity of its environment towards its existence.

III.c. Exterminism as I mentioned above extends to the present—with an aura of life-threatening urgency—the shaping of the population that eugenics projects in a milder manner as a task for the future. Unless that is one expands the range of what is called eugenics so as to include in it exterminism as well. I prefer here keeping a clear difference between the two terms.

IV. Progress Will Prevail

Climate change is likely to reach cataclysmic proportions; air pollution is on the rise, macro- and micro-plastic pollution is pervasive; endocrine disruptors wreak havoc in our reproductive ability and leads to malformation in foetuses and mental handicap; the natural cycles of nitrogen and phosphorus are by now seriously compromised and on their way towards irreversible degradation. These all require as far as the survival of the species is concerned that our overall technological capacities get mobilised in attempts to reverse those lethal trends.

Posthumanism is biased against technology because of its Heideggerian anti-modernist roots. Superhumanism and metahumanism appear to be agnostic on the topic. Transhumanism is of course highly favourably biased towards technological progress as far as human enhancement is concerned through genetics, prosthetics and the less well-defined project of dematerialisation of the self. It would be inconsistent therefore with the overall worldview of transhumanism were it to advocate anti-technological views on other issues.

Space colonisation which falls under this heading is dealt with separately.

V. Survivalism

V.a. The concept of Life in the Underground started historically with bunkers used as bomb shelters. It will have to become a universal feature if—as some scientists forecast for the vicinity of the year 2200—outside temperature will force mammals to an underground life for lack of a sufficiently efficient cooling system on their part, our thermoregulation breaking down at around a constant 40°C temperature.

V.b. Transhumanism holds that rationality has allowed us to realise that we're insufficiently robust, a shortcoming that can now be overcome through genetic enhancement which might provide an answer as well to degraded surroundings as engineering of the genes can ensure that we fit harsher environments.

V.c. Dematerialisation of the self would be the ultimate response to an entirely impaired environment, with toxic gases in the atmosphere, polluted water all around us and a shortage of food. Dematerialised, we wouldn't have to worry about either the quality of the air, water access or the production of assimilable food.

VI. Space Colonisation

Space colonisation would be the appropriate adaptive policy for finding a new virgin environment kind-hearted to our future settlement ventures. Within our own stellar system, in the light of what we know about existing planets and their satellites, colonisation would require deliberately induced *terraforming*.

Space is as such a harsh environment and our current bodies are ill-prepared for life in space. We would need to find ways for countering the menace of cosmic rays and ways for fighting the damaging effects on our body of no gravity or reduced gravity environments.

As well as for technological progress as a whole, transhumanism is among our four anti-humanisms the most favourably inclined towards space exploration and colonisation.

The Cunning of Reason

So much can be said in a cursory review of our four anti-humanisms' attitude towards the risk of extinction.

Our species has a low capacity for reacting adequately to danger of a vast size for two types of reasons, the first being our incapacity of visualising a danger that has never materialised before, the second being our poor intuitive grasp of scale differences, leading us to putting in the same risk bin menaces of different orders of magnitude, tiny and huge, and undervaluing majorly those of the latter type.

As mentioned, our response to the threat of extinction is totally inadequate in size and in vigour. This at least at the conscious level of deliberate decision-making. The question may be raised however if our species, like is the case for many others, doesn't possess a secret survival mechanism that would automatically be triggered when a particular threshold in our capacity for survival is hit.

The notion of the *cunning of Reason* with Hegel refers to these numerous occasions when the human race has made a decisive turn in its history despite no individual human holding a clear representation of the process at work. A convincing illustration of that was offered by Arthur Koestler in his *The Sleepwalkers* (1959) where he described a generation of thinkers painstakingly trying to improve the methods and techniques of astrology, a superstitious divinatory technique while, unbeknownst to them, they were putting together and developing what would be the tools of the never-seen-before innovative scientific methodology.

Let us notice that the notion of a *cunning of Reason* at work at the species' level as a whole is inconsistent with the view held by every one of our four varieties of anti-humanism, as each assumes at its centre an individual in full control of what to do next, where willpower allows to transform clearly enunciated intentions into deeds, the whole process being supposedly seamless, with no major deviation being observed between the intention initially formulated and the result attained in the end.

On the contrary, Hegel's *cunning of Reason* is perfectly consistent with the view of "Freudian metapsychology" (psychoanalysis) where every individual is the seat of its own self-based *cunning of Reason*, that is, deceived as to its own aims, these resulting from a compromise between conscious and unconscious motives.

Let us ask in that perspective of a possible *cunning of Reason*, the following question: would there be subterraneous processes at work taking over as far as the survival of the species is concerned when our deliberate thinking is inadequate or when our economic system (based on individual profit-seeking strategies) sets insurmountable obstacles? In other words, could adaptive behaviour take place for the species as a whole while our individual behaviour would be counter-adaptive—at least in our representation of it?

Think of the quest for immortality for instance or of augmentation or cyborgisation as promoted by transhumanism.

To what practical purpose would well serve human immortality or, more humbly, a thousand-year life-span, and a human body tinkered in such ways as to much less depend on the specificities of life on earth—or having stopped entirely to depend on those with the script of dematerialisation—if not for travelling to neighbouring stars? To that extent, transhumanism in particular, among our four brands of anti-humanisms, could very well be subterraneously and surreptitiously at work in our psyche as the worldview supporting our long-term adaptation to the fate that will inexorably be ours in the times to come: planet Earth's environment having become unliveable to us, leaving as alternatives for the future, machine-successors to us or life for us on other planets.

Let us be proud of where we currently stand, whatever perils are surrounding us at the moment, and of what we have managed to achieve so far, having relied on our own strengths only. We were indeed dishearteningly alone and all we did was in truth left to our own devices, even if we have been imagining all along human history numerous superhuman and supernatural entities supervising our deeds, lurking behind the scenes.

Apart from existential risks of an outer nature such as collision with an asteroid and super-volcanoes or self-engineered risks such as war, the main event likely to prevent us from pursuing on our current survival path is trespassing lethally the boundaries of the carrying capacity of our environment, whatever that environment might be: our current planet or a new abode that we decided to colonise. It is up to us escaping the millenary nervous breakdown wherein we plunge back every time the thought crosses our mind that we are on our own. Let us create an even more extraordinary world, the one that knowledge makes possible. It might be indeed that God created us to its image but having no clue what that image might very well be, the horizon is limitless, and it is ours to define.

References

The "Book of Genesis" in *The Bible*

del Val, J. and S. L. Sorgner, *A* Metahumanist *Manifesto*, 2010, http://metabody.eu/wp-content/uploads/2016/02/A-METAHUMANIST-MANIFESTO.pdf

Fuller, S., and V. Lipinska, *The Proactionary Imperative, a Foundation for Transhumanism*, Basingstoke: Palgrave Macmillan, 2014

Koestler, A., *The Sleepwalkers. A History of Man's Changing Vision of the Universe*, London: Hutchinson, 1959

Nietzsche, F., *The Antichrist. Curse on Christianity* [1888] 1895, New York: Alfred A. Knopf, 1918

Nietzsche, F., *Ecce Homo* [1888] 1908, http://www.nietzschesource.org/#eKGWB/EH

Oreskes, N., and E. M. Conway, *Merchants of Doubt. How a Handful of Scientists Obscured the Truth on Issues from Tobacco Smoke to Global Warming*, New York: Bloomsbury Press, 2010

Ethics and Complexity: Why Standard Ethical Frameworks Cannot Cope with Socio-Technological Change

Clément Vidal and Francis Heylighen

Standard ethical frameworks struggle when dealing with transhumanism, ecological issues and the rise of technodiversity since their focus is guiding and assessing human behaviour. Ethics needs a Copernican revolution so as to be able to deal with all moral agents, encompassing not only humans, but also artificial intelligent agents, robots and organisations of all sizes.

We argue that embracing a complexity worldview is the first step towards this revolution and that standard ethical frameworks are still entrenched in the Newtonian worldview. We first spell out the foundational assumptions of the Newtonian worldview, where all change is reduced to material particles following predetermined trajectories governed by the laws of nature. However, modern theories of physics such as relativity, quantum mechanics, chaos theory and thermodynamics draw a much more confusing and uncertain picture and have inspired indecisive, subjectivist, relativist, nihilistic or postmodern worldviews. Having as

C. Vidal (✉) • F. Heylighen
CLEA, Vrije Universiteit Brussel, Bruxelles, Belgium
e-mail: contact@clemvidal.com; fheyligh@vub.ac.be

P. Jorion (ed.), *Humanism and its Discontents*,
https://doi.org/10.1007/978-3-030-67004-7_12

foundations cybernetics, systems theory and the recent sciences of complexity, we introduce the complexity worldview that sees the world as interactions and emergent organisations arising from them. This new complexity worldview allows us to underline the limitations of standard ethical frameworks such as deontology, theology, consequentialism, virtue ethics, evolutionary ethics and pragmatism.

Introduction

In our accelerating society, everything is changing faster and faster, even change itself (Russell 2017). Human nature is fundamentally affected by these changes. For example, biotechnologies are changing the functioning of our organism, and by interfacing our bodies with technologies, we are challenging our very identity, giving rise to cybernetic organisms, or cyborgs that only existed in the realm of science fiction a few decades ago.

Cognitive technologies also affect human nature by extending the scope and power of the human mind, in much the same way as sensory and motor technologies extend the scope and power of the bodily perceptual abilities and movement (Dror and Harnad 2008, 3). Indeed, language, writing, printing, books and calculators already support our memory and thinking, and the amplification of our mental abilities only accelerates with the web, algorithms or artificial intelligent (AI) agents.

Obviously, any innovation can be used for positive or negative purposes. Therefore, we need an ethics to guide not just human behaviour, but also the applications of technology and transhumanism. Nobody is able to stop technological progress, so it seems the best we can do is to carefully regulate these new developments. But according to which values, rules or criteria?

The challenge of ethics today is to find ways to understand, guide and steer the actions of a single global village of some 8 billion people and many billions of technological artefacts that we design, build, interact with and increasingly depend upon.

Historically, ethical norms, rules and regulations appeared as sophisticated mechanisms to preserve human communities by inhibiting selfish behaviour. The evolutionary origins of morality show that even the most primitive bands of hunter-gatherers already have norms that suppress bullies, non-altruists and cheaters (e.g. Boehm 2012). Boehm further argues that the regulatory mechanisms of public opinion, conflict management and agreements on rules are present at all group sizes around the world: from foraging bands to farming tribes, chiefdoms, kingdoms, early civilisation and modern nations.

Evolutionarily, the formation of large groups of humans is very recent, so we arguably have little instinctive moral sense to deal with our globalised society. Indeed, we evolved to focus on social aspects such as unity, hierarchy, equality and proportionality within a small group (Rai and Fiske 2011). The evolutionary roots of human morality are thus not sufficient to deal with the ethical challenges we are presently facing. Actually, each time the human group size increased significantly, new regulation and control mechanisms had to be designed. In our transforming and complexifying society, the relevant group becomes the entire planet. Therefore, we are facing the challenge of creating regulatory mechanisms at a planetary scale, hopefully towards a *global ethics*.

The issue is delicate because ethics is not just about humans anymore. The rise of AI, robots, and transhumanism leads to the emergence of new moral agents, such as augmented humans, cyborgs, AI agents, robots, and different kinds of autonomous or semi-autonomous devices that are confronted with ethical decisions to make.

Ethics is essentially about evaluating to what extent behaviour is considered good or bad, whether it is by focusing on the rules by which one operates (deontic ethics) or the consequences of one's actions (consequentialist ethics). However, there are intrinsic limits to these approaches, especially as they implicitly assume that the rules are absolute, or the consequences perfectly clear and predictable in a deterministic fashion. Such assumptions make sense in a *Newtonian worldview*, where all change is reduced to material particles following predetermined trajectories governed by the laws of nature. Associated with this is an ontology of *being*, of static objects, with permanent identities, whose only activity is to change their position within an absolute space and time.

This is to be contrasted with an ontology of *becoming*, where identities emerge, change and eventually disappear. This means that the distinctions and categories we use to understand our situation are not fixed, but fluid. This is indeed how we experience our day-to-day, ever more changeful reality. The Newtonian worldview thus lacks the flexibility for dealing with our complex world. Unfortunately, even though the postmodern worldview emphasises fluid identities, unpredictability and uncertainty, in practice this offers merely a second-order critique of the limitations of classical reason rather than a substantive ethical guide.

Additionally, traditional ethical frameworks remain anthropocentric and therefore cannot deal with transhumanism, ecological issues and the rising technodiversity. Ethics needs its Copernican revolution to be able to deal with all moral agents, not only humans. A few authors have explored a more "universal ethics" based on evolution and development (Vidal 2014, chap. 10), thermodynamics (Ostwald 1912; Hammond 2005; Korbitz 2010; Vidal 2014), theoretical computer science (Bennett 2014; Delahaye and Vidal 2019), or by thinking about extra-terrestrial ethics (Vakoch 2014).

We wish to situate this chapter within this "universal ethics" effort, but will here focus on the need to go beyond the Newtonian worldview by criticising traditional ethical frameworks based on it. Our goal in future works is to use universal *principles of cybernetics and system theory* to propose strategies for a systems ethics that is adapted to our volatile, uncertain, complex and ambiguous world.

In section "From a Newtonian to a Complexity Worldview" we trace the roots of western thought in ethics as stemming from a Newtonian, mechanistic worldview in which fixed, clear rules can be used to reach unambiguous, fixed goals in a deterministic and predictable way. We then argue that such a Newtonian worldview is now largely outdated, being insufficient both scientifically and as a model for ethics. We put forth the fundamentals of the complexity science worldview, where there are no absolute rules, no absolute categories and only limited predictability.

In section "Limitations of Traditional Ethical Frameworks" we criticise from this complexity worldview perspective the main existing ethical frameworks, namely deontology, theology, consequentialism, virtue ethics, evolutionary ethics and pragmatism.

From a Newtonian to a Complexity Worldview

Newtonian Science as an Ideal for Ethics

Philosophy has often envied the rigorous method, clarity, dependability, usefulness and progress of mathematics and science (see e.g. Adler 1993). Spinoza's *Ethics* famously uses the axiomatic method of Euclid and applies it to philosophy. However, a great addition to the formal axiomatic method appeared with Newton's publication of the *Principia Mathematica*. The universal law of gravitation he put forth is a landmark in the birth of modern science, where one clear and concise law allows predicting the motion of both terrestrial and celestial bodies.

Newtonian science is thus expressed in terms of *laws* that are absolute and timeless, and that govern the behaviour of all things. The predicted trajectories of objects are certain, objective, time-independent, time-reversible and context-independent. These are desirable features for scientific theories because they allow prediction and control. A theory that works uncertainly or only at a particular time and place and for a particular subject is not useful compared to a theory that is certain and works always and everywhere (Heylighen, Cilliers, and Gershenson 2007).

If we would extrapolate the laws of motion to the behaviour of all things, then Newtonian mechanics implies the philosophical doctrine of *determinism*, where all notions of uncertainty, freedom, surprise, novelty, emergence, creativity and evolution have become empty.

In this extreme form, ethics becomes pointless because everything is determined in advance from initial conditions and universal laws. Looking towards the past, it means that nobody can be held responsible for his or her past actions, because the game was already set. Looking towards the future, there is no point in choosing the best course of actions because everything is already determined, and the only attitude we can hold is fatalism.

This incompatibility has disturbed generations of philosophers who were trying to reconcile the free will we experience as human beings with the determinism of nature implied by the Newtonian worldview.

Complexity Science as an Ideal for Ethics

In our globalised society, the economic, social, technological and ecological systems that we are part of become ever more interdependent. This means that changes in one component can affect any others, often in an unpredictable way, leading to undesirable systemic effects, such as cascading failures (see e.g. Helbing 2013). These complex, uncertain and unpredictable dynamics create stress for individuals and their organisations, in part because their traditional methodologies based on reductionist Newtonian thinking simply do not apply, as they try to isolate phenomena instead of examining their relationships.

The Newtonian worldview has been undermined by five major scientific breakthroughs.

First, the discovery of undecidability in formal systems in logic and mathematics (Gödel 1931) shows that *absolute certainty* in terms of being able to prove the truth of falsity of any proposition does not exist in most formal axiomatic systems. Since physics and most other sciences use mathematics as a language, these fundamental limitations contaminate the whole scientific edifice.

Second, Einstein's discovery of relativity theory has shown that the absolute space-time of Newtonian physics does not hold and that one can only *compare* relative positions, velocities or generally measurements (see Nottale 2019 for a history of relativity theories). One core lesson of relativity theory is that we need to take into account the observer making the measurements.

Third, the development of quantum mechanics has shown that Newtonian mechanics fails when it is applied to atoms and elementary particles. Two fundamental conclusions are the *intrinsic limit on predictability* implied by the Heisenberg uncertainty principle, and the fact that the act of observing particles influences the results of experiments, undermining the scientific ideal of independent observation and objectivity.

Fourth, chaos theory has shown that many systems behave nonlinearly, meaning that there is no proportionality between cause and effect. For example, microscopic fluctuations in initial conditions may be greatly amplified, a phenomenon known as the "butterfly effect", because it implies that the fluttering of a butterfly can produce a hurricane. The lesson here is also *unpredictability*, although a practical one, because chaos is expressed in terms of deterministic steps, yet the non-linearity can step-by-step transform unobservably small fluctuations in huge macroscopic effects.

Fifth, thermodynamics challenges Newtonian science by requiring a statistical description of the world, which makes predictability less absolute. Moreover, the second law of thermodynamics shows that disorder can only increase or stay the same in a closed system, but never decrease. This implies a fundamental *irreversibility*, which defines an *arrow of time* that is absent in classical and relativistic mechanics.

Relativity, quantum mechanics, chaos theory and thermodynamics define contemporary physics. However, they are each counterintuitive theories that are difficult to reconcile and integrate in an elegant, coherent picture such as the Newtonian worldview. Rather, it is tempting to lapse into an indecisive, subjectivist, relativist or nihilist worldview if one focuses on all the limitations that these theories have put to the fore.

However, modern physical theories still have fundamental limitations in that they do not deal with goal-directed agents, which are of central importance for ethics. As such, physics will always remain an insufficient source of inspiration for ethics. One has to go beyond purely physical theories to find relevant scientific concepts and models to understand purposeful systems such as animals, humans, organisations, societies or our ever more autonomous technological artefacts.

We will now argue that evolution, cybernetics, systems theory and complexity sciences provide a promising integrative scientific framework to replace the Newtonian worldview.

In classical physics, there is fundamentally nothing new that can emerge, no creativity. This is because the basic structure of classical causality *conserves distinctions* (Heylighen 1989): equal causes have equal effects, and distinct causes have distinct effects. By contrast, the evolutionary process is essentially creative and as such *creates distinctions*: the same initial form can give rise to an ever-larger diversity of variations. Evolution is what produces goal-directed organisms, and its study is thus fundamental to descriptive ethics, and to a certain extent, to prescriptive ethics too.

The overall effect of the myriad interactions, mutations, recombinations, non-linear dynamics and so forth, even if each could be formalised as an explicit rule, would not be predictable. It would seem extremely hard if not impossible to predict the present bio- and technodiversity starting from the first life-forms that appeared 4 billion years ago. This is why scientists today make use of computer simulations to run and re-run models step by step and try to find general patterns (in biology, see e.g. Bedau 2009; in cosmology, see Vidal 2013; in computer science, see e.g. Wolfram 2002).

Complexity science can be seen as the continuation of cybernetics and systems theory (for some history, see Von Bertalanffy 1972; Heylighen and Joslyn 2001; Pickering 2010). The complexity worldview sees the *world as interactions and their emergent organisations*, instead of seeing the world as separate particles and forces. This leads to efforts to model systems, their organisations, interrelations, coupling and more, and is a major shift from the materialist, mechanistic, analytical, reductionist approach of Newtonian science (Laszlo 1972a, 1972b). Thus, complexity thinking emphasises processes over static objects, relations over isolated elements, context-dependence over universality, flow over equilibrium and wholes over parts.

This systems perspective allows us to formulate the problem of ethics in a very general way. The fundamental issue follows from the *principle of suboptimisation* (Machol 1965): what is optimal for a component or subsystem is in general not optimal for the system as a whole. For example, for a cell or tissue it is best to be able to grow without restraint, but for the system as a whole such unbridled growth behaves like a tumour that may endanger the whole organism. This principle is a corollary of the more general principle of *emergence*, which notes that a coherent whole has emergent properties that cannot be found in the parts. Indeed, if there were no emergence, what would be best for the parts would also be best for the whole.

A classical application to ethics is the prisoner's dilemma (Axelrod 1984), where the best option for one individual is the worst for the other, while the optimal, cooperative solution for both individuals is not achieved because the temptation to selfishly defect is too high. A similar ethical conundrum is the "free-rider" problem: for an individual it is best to profit from a collective arrangement such as the public transport system, but without putting in any resources, for example, by not paying. However, if everyone would behave like that, the collective system would collapse, and everyone would suffer. Thus, the task of ethics could be defined as the formulation of rules or guidelines that reconcile the values of the whole and its different parts, so that all components work together synergetically, rather than being in conflict or opposition.

Limitations of Traditional Ethical Frameworks

We will now argue that traditional ethical frameworks are based implicitly on the Newtonian worldview, and thus suffer from serious limitations. First, one can note that ethical theories emphasise *objectivity*, but when one views ethical issues from an objective standpoint, one fails to address the real difficulties of ethical situations that will always occur from the *subjective* point of view of an individual agent confronted with the situation. As such, ethical theories are of little help in guiding actions in the midst of a real ethical dilemma. Ethics has been criticised for taking the easy option of looking down from a detached, "objective" and judgemental position rather than getting engaged in the midst of the action (see e.g. Hampshire 1949; Varela 1999; Whitbeck 2011; Sweeting 2018).

Deontology

Deontology is an ethical framework that assumes that there are unambiguous and time-independent rules, comparable to universal laws in physics. In a very stable world, with few interactions or changes, such rules might work reasonably well. This is clearly not the case of today's world with its complex and intertwined ethical issues. The *context-dependence* of ethical issues makes it unclear by which rule (or maxim) one should act. Additionally, rules may change or may become irrelevant and no longer applicable as the world changes.

It is also unclear whether any maxim could be universalised, because each situation is unique. Let us criticise Kant's famous "categorical imperative" from this perspective. It states that you should always act in such a manner that you would wish this manner of acting to become a universal rule, applied by everybody everywhere. The imperative is a powerful way to reflect on the consequences of actions, because imagining the universalisation is fundamentally a two-step process: first a change of perspective to imagine what would happen if everybody was following the same rule, and second, going back to choose an action from one's own perspective. For example, if one would follow the maxim "do not hesitate to lie", the social fabric of trust would quickly fall apart.

Unfortunately, it is totally beyond human cognitive abilities to imagine what would or would not hold for all humans, in any situation, at any given time. In particular in the case of technology ethics, it is even harder to predict or imagine what effects the application of some new technology would have. In practice, there will always be exceptional situations where the generalisation would not hold, situations that one cannot even imagine.

Theology

Fixed rules can be useful in a given stable context in space and time. Religions are structures that provide rule-based guidance, such as the Ten Commandments or the Jewish/Islamic injunction to not eat pork. Although the rule against eating pork may have been useful to avoid diseases carried by pigs in the historical context in which the rule was formulated, with today's high standards of hygiene, eating pork is perfectly safe. Therefore, it is for purely traditional reasons, and not anymore for health reasons that this rule is still observed in some religions. Similarly, at least half of the Ten Commandments (such as the commandment that forbids making idols) are no longer considered as moral or legal obligations in a modern, secular society.

This observation illustrates that we should be extremely wary to extrapolate moral rules, following the Newtonian example, to *any* situation, *any* context, *for all* people, and *all* space and time.

Of course, religions are much more than a set of rules. Religions are cultural systems that morally orient and socially bind human groups through historically extensive networks of ritual practices, beliefs, values, institutions and narratives (Hogue 2010). In practice, religions do not follow their millennia-old rules algorithmically, but apply them flexibly while providing adaptive mechanisms such as conversation during confessions, community support, dialogues with religious authorities or wise people. This adaptivity is necessary to guide action in tricky situations. Its existence in part explains the continuing success and relevance of religions even in an age of science.

Consequentialism

Consequentialist ethics proposes that we should choose actions based on their practical consequences rather than on a priori rules. As such, it may seem more appropriate to the messiness of the real world. However, consequentialism implicitly assumes that one can unambiguously deduce the consequences of actions, just as one can predict the future positions of planets using Newtonian mechanics. Of course, things are not as easy as that! Perfect predictability in science is only possible in narrowly defined, specific cases. In the psychological or sociological realms where ethical behaviour is situated, one can only predict statistically at best.

Consequentialism also fails at taking into account creativity, emergent features or context-dependence. It states that we should choose the action that maximises some overall criterion of goodness or benefit, such as "the greatest happiness for the greatest number". This assumes that we can formulate some clear, unambiguous and unchanging measure of benefit—which in mathematical models of decision-making is called a "utility function" (see also Vodonick 2016). One might somewhat generalise this requirement by considering multiple criteria at multiple levels, but even this would fail to take into account the fact that goals or criteria can become irrelevant during a course of action. Also multiple criteria optimisation leads to different equally good solutions, so one could not easily decide between different ways to satisfy many criteria.

Additionally, there is a fundamental impossibility in the consequentialist dream of mapping and evaluating all possible trees of possible actions, because this would be impossible to compute, as it would quickly exhaust all the computational resources of the universe (i.e. the Bremermann limit, see Ashby 1968).

Interestingly both deontology and consequentialism emphasise different aspects of the same Newtonian worldview: deontology emphasises laws, while consequentialism emphasises predictable consequences from such laws. But their underlying worldview remains the same Newtonian one.

Virtue Ethics

Virtue ethics is the philosophy that we should maximally try to develop good qualities (virtues) in ourselves, such as wisdom, empathy and moderation, while minimising negative qualities (vices), such as jealousy, stinginess or prejudice. As Sweeting (2018, 14) noticed, virtue ethics is one of the ethical frameworks most compatible with cybernetics and the complexity worldview. As a matter of fact, Chambers (2001) has argued that Aristotle's Doctrine of the Mean, where each virtue is considered a virtuous mean along a spectrum between two opposite poles (e.g. courage is in between cowardice and foolhardiness), can be seen as a learning process driven by trials and errors and negative feedback until the right virtue is acquired.

The limitation of virtue ethics is that it is focused on making virtuous *people*, not virtuous organisations, technologies or systems. In a subsequent paper, we plan to explore more universal, cybernetic virtues that could apply to any moral agent.

Also, even if virtue ethics helps to make better people, it does not help us to make a good decision in a real ethical dilemma involving a choice between incompatible virtues.

Evolutionary Ethics

Evolutionary ethics is the position that our moral rules and values have their roots in human evolution, which has given us instincts to recognise "fit" behaviour that promotes the survival of the group in the long term and "unfit" behaviour that is detrimental to this. Evolutionary ethics has a bad reputation among social scientists mostly because it tends to be associated with the outdated ideology of social Darwinism. However, as argued by David Sloan Wilson (Wilson 2019, chap. 1), this rests on a misunderstanding:

> If grave social injustices were committed in Darwin's name, such as withholding welfare from the poor, forced sterilisations, racial discrimination, and outright genocide, then this is truly dangerous territory. But this rendering of social Darwinism is largely a myth and the true history is far more interesting and complex. Darwin's theory, properly understood, is centred on cooperation, and Darwin and others were clear about this from the beginning.

A desideratum of an ethical theory is that it can apply to the *past*, to judge or justify past courses of actions, as well as to the *future*, to choose a future course of action. Another common misunderstanding about evolutionary ethics is that it could be used to justify primitive selfish behaviour, but this is not the case. For example, an attorney arguing in court the following would not gain much traction: "It's true that my client raped this young girl, but really it came from his biological reproductive instinct, so he should be forgiven".

Descriptive evolutionary ethics as developed in the field of moral psychology has made much progress and has led to deep insights about where our morality comes from and why we have such or such moral instincts (see e.g. Rai and Fiske 2011; Graham et al. 2013). However useful and insightful, such descriptive ethics does not tell us what to do and how to act. It is a scientific, objective, detached way to study morality, rather than a normative ethics that can guide action effectively.

Can or should we use evolutionary ethics normatively? This is not straightforward. The first answer appears to be "no" because the adaptive solutions found by past biological or cultural evolutionary wisdom are adaptations to biological or sociological ecologies *at that time in the past*. There is no guarantee that they would still prove adaptive in present or future situations (Campbell 1979).

So one has to use evolutionary wisdom wisely, as knowledge that may help us to design solutions to complex problems (ethical or others) by applying evolutionary mechanisms that have shown to be successful, however, while being aware that, like in the stock market, past success does not guarantee future performance.

A general inspiration for an evolutionary ethics that is more future-oriented is the postulated direction of evolution towards increasing complexity and adaptivity (e.g. Dawkins 2006; 2003, sec. 5.4; Stewart 2000; Heylighen 1999; Vidal 2014, 286–89). Most evolutionary theorists know that evolution cannot move in any direction, and is constrained in many ways. However, the topic remains controversial even today because publications about the direction of evolution were initially banned by evolutionary scientists to build their discipline and profession (Ruse 1996).

Of course, we should be wary of the spectre of the naturalistic fallacy (you cannot derive "ought", i.e. ethical injunctions, from "is", i.e. facts about evolution). Yet, we can note that the issue is analogous to the problem of induction in epistemology (Hume 2005). Just like induction cannot lead to absolute truths in science (Popper 1959), induction cannot lead to fundamental moral laws in ethics. Nevertheless, the discovery of the impossibility to fully validate inductive reasoning in science did not stop science from progressing through induction. Therefore, the discovery of the impossibility to fully validate moral reasoning does not preclude ethical naturalism (see also Curry 2006). The problem of induction and the naturalistic fallacy are salient only if one holds an epistemic realist or a moral realist position, that is, the assumption that scientific theories and moral rules reflect a reality independent of the observer. In a constructivist or cybernetic vision of knowledge and morality, scientific models and moral laws are both constructions, so these critiques do not hold.

Pragmatically, if one can find or design solutions to difficult ethical issues inspired by evolution rather by alternative sources, then evolutionary ethics is worth using.

Pragmatic Ethics

Pragmatic ethics states that our moral rules and values need to be tested out and if necessary revised in practical situations. If a rule consistently produces undesired effects, then we should replace it by a better rule, thus continuously improving and updating our ethical system. This is similar to how science continuously improves its theories through reasoning, observation and experiment. Pragmatic ethics is arguably the most flexible framework to deal with ethical issues, because it has no fixed rules or values. It is thus naturally context-dependent, fluid, open to changing situations and ready to transform itself.

Note that this "rule of self-improvement" is a meta-rule, so there is no contradiction here. Knowing the history of ethical principles and rules could certainly help to first realise that rules, norms, principles *did change* throughout history, as we saw with the example of the injunction against eating pork. If we apply this remark to the future, this suggests improving and possibly going beyond any existing rule.

However, the question of how, when and why to improve on moral rules or strategies remains difficult and open to debate. An analogy with scientific progress is to keep moral rules that "work best" and reject or modify the ones that do not. However, the analogy with science is not straightforward. How should we perform "ethical experimentation"? Would it even be ethical to conduct such experiments? It also leaves open the question what we mean by "working best"? When do we decide that moral rules are unfit and should be rejected? What are the selection criteria for ethical rules, norms or theories?

One could also object that pragmatism is rather a bottom-up approach, which seeks to solve problems as they appear. As such, it seems to lack top-down, universal or overarching values. This would lead pragmatists to care mostly for a short-sighted time horizon.

Conclusion

The Newtonian worldview is so deeply entrenched in any rational endeavour that it is easy to overlook how constraining its influence is, even for a domain seemingly far removed from Newtonian mechanics like ethics. We have spelled out its fundamental assumptions, and shown how they have been largely refuted by modern physics. Unfortunately, ethical frameworks have become disconnected from modern science, ignoring in particular the lessons of the new science of complexity. This makes them largely inapplicable to our fast evolving socio-technological society, in which developments such as transhumanism, autonomous AI and cyborgisations open up a wide range of novel ethical issues. Just as it takes years for school textbooks to become up-to-date with current scientific knowledge, ethical frameworks are largely out of date with modern physics and the complexity worldview.

A common reaction to the rigidity and absolutism of the Newtonian worldview is to fall into the other extreme and endorse a postmodern, relativistic philosophy in which anything goes. This is not our stance. In this chapter we criticised the basic assumptions of the standard ethical frameworks from the complexity perspective. In a subsequent paper, we also plan to propose methods, principles and strategies from cybernetics, system theory and complexity science that may provide foundations for a new ethics. Our aim will be to discuss constructive strategies for dealing with wicked, changeful and complex problems that may offer guidance to any moral agent, whether it is an individual, an AI agent or an organisation.

Acknowledgements Clément Vidal thanks Ben Sweeting for insightful conversations.

References

Adler, Mortimer J. 1993. *The Four Dimensions of Philosophy*. Macmillan USA.
Ashby, W. Ross. 1968. "Some Consequences of Bremermann's Limit for Information-Processing Systems." *Cybernetic Problems in Bionics*, 69–76.

Axelrod, Robert M. 1984. *The Evolution of Cooperation*. New York: Basic Books.

Bedau, Mark A. 2009. "The Evolution of Complexity." In *Mapping the Future of Biology*, 266:111–30. Boston Studies in the Philosophy of Science. Springer Netherlands.

Bennett, C. H. 2014. *Evidence, Computation, and Ethics*. Simons Symposium on Evidence in the Natural Sciences.

Boehm, Christopher. 2012. *Moral Origins: The Evolution of Virtue, Altruism, and Shame*. New York: Basic Books.

Campbell, Donald T. 1979. "Comments on the Sociobiology of Ethics and Moralizing." *Behavioral Science* 24 (1): 37–45. doi:https://doi.org/10.1002/bs.3830240106.

Chambers, Jean. 2001. "A Cybernetic Theory of Morality and Moral Autonomy." *Science and Engineering Ethics* 7 (2): 177–92. doi:https://doi.org/10.1007/s11948-001-0039-1.

Curry, Oliver. 2006. "Who's Afraid of the Naturalistic Fallacy?" *Evolutionary Psychology* 4 (1): 147470490600400130.

Dawkins, Richard. 2003. *A Devil's Chaplain: Reflections on Hope, Lies, Science, and Love*. Boston: Houghton Mifflin Co.

Dawkins, Richard. 2006. *The Selfish Gene*. Oxford University Press.

Delahaye, J. P., and C. Vidal. 2019. "Universal Ethics: Organized Complexity as an Intrinsic Value." In *Evolution, Development and Complexity: Multiscale Evolutionary Models of Complex Adaptive Systems*, edited by Georgi Yordanov Georgiev, Claudio Flores Martinez, Michael E. Price, and John M. Smart. Springer. doi:https://doi.org/10.5281/zenodo.1172976. https://zenodo.org/record/1285656.

Dror, Itiel E., and Stevan Harnad. 2008. "Offloading Cognition onto Cognitive Technology." In *Cognition Distributed: How Cognitive Technology Extends Our Minds*, 1–23. http://arxiv.org/abs/0808.3569.

Gödel, Kurt. 1931. "Über Formal Unentscheidbare Sätze Der Principia Mathematica Und Verwandter Systeme I." *Monatshefte Für Mathematik Und Physik* 38 (1): 173–198.

Graham, Jesse, Jonathan Haidt, Sena Koleva, Matt Motyl, Ravi Iyer, Sean P. Wojcik, and Peter H. Ditto. 2013. "Moral Foundations Theory: The Pragmatic Validity of Moral Pluralism." In *Advances in Experimental Social Psychology*, edited by Patricia Devine and Ashby Plant, 47:55–130. Academic Press. doi:https://doi.org/10.1016/B978-0-12-407236-7.00002-4. http://citeseerx.ist.psu.edu/viewdoc/download?doi=10.1.1.693.848&rep=rep1&type=pdf.

Hammond, Dick. 2005. *Human System from Entropy to Ethics*. Eulogy Edition.
Hampshire, S. 1949. "Fallacies in Moral Philosophy." *Mind* 58 (232): 466–82. https://www.jstor.org/stable/2250876.
Helbing, Dirk. 2013. "Globally Networked Risks and How to Respond." *Nature* 497 (7447): 51–59. doi:https://doi.org/10.1038/nature12047.
Heylighen, F. 1989. "Causality as Distinction Conservation: A Theory of Predictability, Reversibility and Time Order." *Cybernetics and Systems*, no. 20: 361–84. http://www.informaworld.com/index/770570830.pdf.
Heylighen, F. 1999. "The Growth of Structural and Functional Complexity During Evolution." In *The Evolution of Complexity: The Violet Book of "Einstein Meets Magritte*, edited by F. Heylighen, Johan Bollen, and A. Riegler, 17–44. Dordrecht: Kluwer Academic Publishers. http://pespmc1.vub.ac.be/Papers/CausalityDist.pdf.
Heylighen, F., P. Cilliers, and C. Gershenson. 2007. "Complexity and Philosophy." In *Complexity, Science and Society*, edited by Jan Bogg and Robert Geyer, 117–34. Oxford: Routledge. https://arxiv.org/abs/cs/0604072.
Heylighen, F., and C. Joslyn. 2001. "Cybernetics and Second Order Cybernetics." *Encyclopedia of Physical Science & Technology* 4: 155–70. http://pespmc1.vub.ac.be/Papers/Cybernetics-EPST.pdf.
Hogue, Michael S. 2010. *The Promise of Religious Naturalism*. Rowman & Littlefield Publishers.
Hume, David. 2005. *A Treatise of Human Nature*. New York: Barnes & Noble.
Korbitz, Adam. 2010. "Metalaw and SETI: Xenology, Metalaw and Thermoethics." *Metalaw and SETI*. December 2. http://web.archive.org/web/20110708053037/http://metalawandseti.blogspot.com/2010/12/xenology-metalaw-and-thermoethics.html.
Laszlo, E. 1972a. *Introduction to Systems Philosophy – Toward a New Paradigm of Contemporary Thought*. Gordon & Breach Science Pub.
Laszlo, E. 1972b. "Systems Philosophy: A Symposium. The Case for Systems Philosophy." *Metaphilosophy* 3 (2): 123–41. doi:https://doi.org/10.1111/j.1467-9973.1972.tb00045.x.
Machol, Robert Engel, Wilson P. Tanner, and Samuel N. Alexander, eds. 1965. *System Engineering Handbook*. New York: McGraw-Hill.
Nottale, Laurent. 2019. *The Relativity of All Things: Beyond Spacetime*. Translated by Mia Chen. Persistent Press.
Ostwald, Wilhelm. 1912. *Der Energetische Imperativ*. Leipzig: Akademische verlagsgesellschaft m. b. h. http://archive.org/details/derenergetische00ostwgoog.

Pickering, Andrew. 2010. *The Cybernetic Brain: Sketches of Another Future*. Chicago; London: University of Chicago Press.

Popper, K. R. 1959. *The Logic of Scientific Discovery*. Routledge. http://archive.org/details/karlpopperlogicofscientificdiscovery_61_r.

Rai, Tage Shakti, and Alan Page Fiske. 2011. "Moral Psychology Is Relationship Regulation: Moral Motives for Unity, Hierarchy, Equality, and Proportionality." *Psychological Review* 118 (1): 57–75. doi:https://doi.org/10.1037/a0021867. https://cloudfront.escholarship.org/dist/prd/content/qt8pn9p2n7/qt8pn9p2n7.pdf.

Ruse, Michael. 1996. *Monad to Man: The Concept of Progress in Evolutionary Biology*. Cambridge, Mass: Harvard University Press.

Russell, P. 2017. "Blind Spot: The Unforeseen End of Accelerating Change." *Spanda Journal* VII (1): 31–42.

Stewart, John. 2000. *Evolution's Arrow: The Direction of Evolution and the Future of Humanity*. Chapman Press. http://users.tpg.com.au/users/jes999/EvAr.pdf.

Sweeting, Ben. 2018. "Wicked Problems in Design and Ethics." In *Systemic Design: Theory, Methods, and Practice*, edited by Peter Jones and Kyoichi Kijima, 119–43. Translational Systems Sciences. Tokyo: Springer Japan. doi:https://doi.org/10.1007/978-4-431-55639-8_5. https://philpapers.org/rec/SWEWPI.

Vakoch, Douglas A., ed. 2014. *Extraterrestrial Altruism: Evolution and Ethics in the Cosmos*. The Frontiers Collection. Springer Berlin Heidelberg.

Varela, Francisco J. 1999. *Ethical Know-How: Action, Wisdom, and Cognition*. 1 edition. Stanford, Calif: Stanford University Press.

Vidal, C. 2013. "Artificial Cosmogenesis: A New Kind of Cosmology." In *Irreducibility and Computational Equivalence: 10 Years After A New Kind of Science*, edited by H. Zenil, 157–82. Heidelberg: Springer. doi:https://doi.org/10.1007/978-3-642-35482-3_13. http://arxiv.org/abs/1205.1407.

Vidal, C. 2014. *The Beginning and the End: The Meaning of Life in a Cosmological Perspective*. New York: Springer. http://arxiv.org/abs/1301.1648.

Vodonick, John. 2016. "Pragmatism-Beyond Epistemology: An Ethical Approach to Systems Decision Process." *Journal of Organisational Transformation & Social Change* 13 (1): 43–53. doi:https://doi.org/10.1080/14779633.2016.1148900.

Von Bertalanffy, Ludwig. 1972. "The History and Status of General Systems Theory." *Academy of Management Journal* 15 (4): 407–26. doi:https://doi.org/10.5465/255139.

Whitbeck, Caroline. 2011. *Ethics in Engineering Practice and Research*. 2 edition. Cambridge University Press.
Wilson, David Sloan. 2019. *This View of Life: Completing the Darwinian Revolution*. New York: Pantheon.
Wolfram, S. 2002. *A New Kind of Science*. Wolfram Media Inc., Champaign, IL.

Index[1]

[1] Note: Page numbers followed by 'n' refer to notes.

P. Jorion (ed.), *Humanism and its Discontents*,
https://doi.org/10.1007/978-3-030-67004-7

Papier | Fördert gute Waldnutzung
FSC® C083411

Kolibri 360 GmbH
Ferdinand-Jühlke-Straße 7
99095 Erfurt, Deutschland
produktsicherheit@kolibri360.de